CIRIA C756

CDM 2015 – workplace 'in-use' guidance for designers, second edition

A Gilbertson

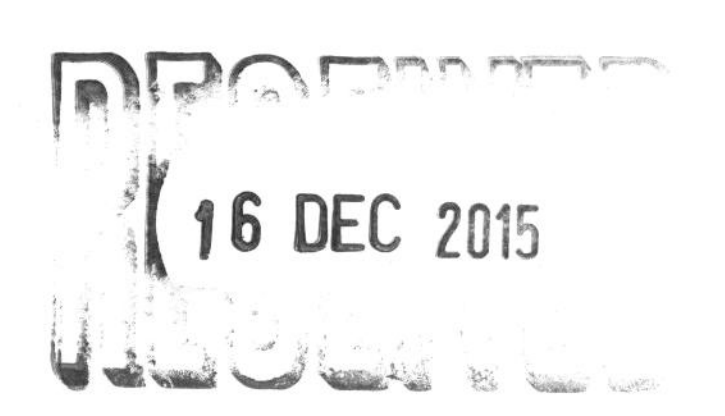

ciria
Griffin Court, 15 Long Lane, London, EC1A 9PN
Tel: 020 7549 3300 Fax: 020 7549 3349
Email: enquiries@ciria.org Website: www.ciria.org

Summary

The Construction (Design and Management) Regulations 2015 (CDM 2015) apply to construction work undertaken in GB. This guide helps any person or organisation acting as a designer to meet their obligations with respect to designing, taking account of workplace 'in-use' risks. Advice on considering construction risks is provided in CIRIA C755, which also contains more general information about designers' duties under CDM.

Each of the legal duties is explained and advice given on how they may be effectively discharged. This new edition takes account of the Health and Safety Executive (HSE) guidance *Managing health and safety in construction*, L153 (HSE, 2015a). In addition to giving advice on a designer's duties, this guide explains how the work may be carried out in an effective, proportionate manner and provides guidance on the difficult issues that arise in practice.

CDM 2015 – workplace 'in-use' guidance for designers, second edition

Gilbertson, A

CIRIA

C756 © CIRIA 2015 RP1024 ISBN: 978-0-86017-762-3

First published as C663

British Library Cataloguing in Publication Data

A catalogue record is available for this book from the British Library

Keywords		
Health and safety, health and safety in-use, construction management, Construction (Design and Management) Regulations, CDM, CDM 2015, construction work, demolition, design, designers, hazards, maintenance, risk, workplace		
Reader interest	**Classification**	
Construction industry designers, architects, civil engineers, structural, engineers, services engineers, surveyors, consultants, local authorities, facility managers	Availability	Unrestricted
	Content	Original research
	Status	Committee guided
	User	Construction sector designers, (engineers, architects, surveyors, project managers, principal designers), construction industry professionals

Published by CIRIA, Griffin Court, 15 Long Lane, London, EC1A 9PN

About this guide

The Construction (Design and Management) Regulations 2015 (known as 'CDM 2015') affect all construction work. The Regulations place duties upon all designers and this guide is designed to assist in fulfilling those duties with respect to structures used as workplaces, also described in the industry as 'in-use'.

CDM 2015 builds upon earlier health and safety legislation by imposing a framework of duties so that all the parties to a construction project must consider health and safety. The Regulations have an accompanying guidance document, L153 *Guidance on the Construction (Design and Management) Regulations 2015* (HSE, 2015a) which can be accessed via the HSE website. L153 provides guidance from the HSE as to their expectations of how CDM 2015 should be implemented.

Guidance has also been prepared by a CONIAC working group and may be freely accessed through the Construction Industry Training Board (CITB) at: **www.citb.co.uk**

This guide addresses workplace 'in-use' risks and has a companion publication, CIRIA C755, which addresses construction risks, including maintenance.

READERSHIP

This guidance has been produced for designers wishing to develop a full understanding of their function with respect to relevant workplace 'in-use' risks that need to be considered under CDM. It focuses upon the management of risk.

The word 'designer' is a defined term under CDM 2015 and has a broad meaning, going beyond the traditional definition, to include anyone who in the course or furtherance of a business modifies, arranges for or instructs others under their control to carry out design.

Each designer will have his or her own area of decision-making, which will affect risks 'in-use'. Designers need to concentrate on the decisions that they can influence while being aware of the concerns of other designers, which may be affected by their actions.

A general understanding of CDM 2015 and its guidance document L153 is assumed. For sources of further information, see Appendix A3. In particular, when considering construction risks, designers may refer to CIRIA C755.

Principal designers should also be aware of the information provided in this guidance.

KEY POINTS

Anyone who acts as a designer as defined by CDM 2015 (including a client who imposes specific requirements), has designer duties under CDM.

> All those who contribute to design decisions affecting subsequent use of a structure must consider the risks involved. This requires an understanding of workplace 'in-use' conditions, including related facilities, management activities and the types of accidents and health issues that can occur.

This guide provides a wealth of accessible information about health and safety issues in the use of a wide range of structures. The consideration of these matters must be an integral part of the design decision process. Each designer has a part to play, working as a member of the design team who must make decisions related to health and safety in a co-ordinated manner, liaising with the principal designer (or PD) who has specific legal responsibilities under the regulations in respect of health and safety issues. This guide is not, however, a textbook, and rather it provides prompts to make a designer think and apply their knowledge.

The object of CDM is to embed health and safety management into projects. All designers have a role to play and must provide information about residual health and safety risks for those who operate workplaces, as they plan and manage work.

A designer's duties under CDM require common sense and openness in order to relate to the other duty-holders in a constructive manner. Duties should be carried out in a way that is proportionate, bearing in mind the type of project and the likely level of risk.

Information prepared by designers for the purposes of CDM must:

- focus on health and safety information which competent people would not reasonably anticipate
- be specific to the project
- reflect the level of risk and complexity
- be concise.

> Particular care is needed to prepare concise, focused documentation that is relevant to the project in hand. In this way, it will be of immediate use to the people who need to use it. To achieve this, the use of lengthy, standardised or off-the-shelf, catch-all documents should be avoided. The aim is to produce relevant information that is proportionate to the project and its risks and is useful and effective.

How to use this guide

The guide has been structured so that it can be read from cover-to-cover or consulted by those who simply need to dip in for specific information. The guidance is designed to provoke thought and to improve understanding and knowledge, but it cannot provide complete information for every circumstance. As with all checklist guidance, the user should ask "is there anything else, for this particular project?"

Chapter 1 Introduction

This introduces the requirements of CDM 2015 and examines the issues that arise.

Chapter 2 Information about workplaces

CDM 2015 requires designers to consider the use of structures as workplaces, as defined by regulation, so when carrying out their designs they need to consider the requirements that the Workplace (Health, Safety and Welfare) Regulations 1992 (the 'Workplace Regulations') place upon those who will manage health and safety in the use of a structure designed as a workplace.

Chapter 3 Information about risks

In this guide, a range of typical risks is examined, providing prompts to designers about the issues involved and suggesting key references where further information may be found. Space is provided for users to add their own notes and references.

Chapter 4 Information about typical workplaces

A small selection of typical workplaces is also examined, showing the types of underlying issues that need to be considered when examining 'in-use' risks.

Appendices

A risk checklist is provided, which may assist designers, and an example of its use is shown. Also, further sources of information are referenced.

Acknowledgements

This publication is the result of CIRIA Research Project (RP) 1024. It was written by Alan Gilbertson of Gilbertson Consultants Limited under contract to CIRIA.

Project steering group

The preparation of this publication was guided by a project steering group (PSG) established by CIRIA to advise on technical issues. CIRIA and Gilbertson Consultants Limited would like to express their thanks and appreciation to all members of the PSG and their organisations, which comprised:

Philip Baker	BPS Solutions
Paul Bussey	Scott Brownrigg
John Carpenter	Consultant
Bob Keenan	Sheppard Robson
David Lambert	Kier Group
David Watson	Consultant

CIRIA Project managers

Kieran Tully	Project director
Lee Kelly	Assistant project manager
Clare Drake	Editor

Contents

Examples

Figures

Tables

1 Introduction

1.1 CDM 2015 AND THIS GUIDE

This guide focuses upon the particular issue of 'in-use' risks, ie the risks that may exist for those actually using a structure once it has been constructed. Reference may be made to CIRIA C755 *CDM 2015 – construction work sector guidance for designers, fourth edition*, which provides detailed information about CDM 2015 as it applies to designers.

One particular aspect of CDM 2015 is the use of the term 'risk'. This guidance mainly uses this term to encompass both hazard (the potential for harm) and risk (a measure of the likelihood and the severity of harm occurring). This is because in CDM 2015 the use of language has changed from the earlier CDM regulations in which the words 'hazard' and 'risk' were both used.

This guide should be of interest to principal designers as well as designers. The role of the principal designer and the relationship between designers and principal designers are explored in C755.

1.2 WHAT THE REGULATIONS REQUIRE

Regulation 9 states that (with bold and underlining added by the author of this guide):

(2) *When preparing or modifying a design the designer must take into account the general principles of prevention and any pre-construction information to eliminate so far as is reasonably practicable, foreseeable risks to the health and safety of any person –*
  (a) *carrying out or liable to be affected by construction work*
  (b) *maintaining or cleaning a structure*
  (c) using a structure designed as a workplace.

(3) *If it is not possible to eliminate these risks, the designer must, so far as is reasonably practical:*
  (a) *take steps to reduce or, if that is not possible, control the risks through the subsequent design process*
  (b) *provide information about those risks to the principal designer; and*
  (c) *ensure appropriate information is included in the health and safety file*

(4) *The designer must take all reasonable steps to provide, with his design, sufficient information about the design, construction or maintenance of the structure to adequately assist the client, other designers and contractors to comply with their duties under these Regulations*

In paragraph 82 of the HSE accompanying guidance L153:

> *"Designs prepared for places of work also need to comply with the Workplace (Health, Safety and Welfare) Regulations 1992 (the Workplace Regulations), so designs need to take account of factors such as lighting and the layout of traffic routes."*

It is essential to understand that the definition of a 'structure' is far reaching and includes buildings, services etc. More information is given in CDM 2015 Regulation 2.

CDM 2015 does not specifically require the consideration of 'in-use' risks where the structure is not a workplace, but designers should be aware that under the Health and Safety Act 1974 they are required to consider the consequences of their work upon the health and safety of others, which includes people using a structure that is not a workplace.

1.3 HOW THE APPROACH REQUIRED BY THE CDM REGULATIONS RELATES TO PREVIOUS PRACTICE

There is nothing intrinsically new about the consideration of hazard and risk in the workplace. Those responsible for workplaces have for many years been required to manage risks to workers and others who might come into the workplace. When a new workplace is handed over to those responsible for its management, one of their first activities must be to decide how they are going to manage risk and put actions in place, as required by the Management of Health and Safety at Work Regulations 1999 and the Workplace Regulations 1992. This activity is specific to a particular workplace and relies upon the particular knowledge of the managers of the workplace.

Before the CDM 'in-use' requirements were introduced in 2007 designers normally carried out 'in-use risk' assessment for a future workplace in an informal manner, for example (where relevant) by relying upon compliance with standards laid down by others, such as:

- **authorities** (Building Regulations, the Equality Act and specific regulations applying to particular industry workplaces)
- **clients** (such as the NHS and MoD) who set their own requirements
- **industry bodies** (such as the British Council for Offices), who formulate standards for work sectors
- **technical bodies** (such as CIBSE), who set technical standards.

This guidance tended to be generic and did not always address the specific detailed requirements covered by the Workplace Regulations. Often there was little formal communication on risk between designers and those taking occupation. However, the health and safety (H&S) file required by CDM since 1995 should, where available, contain useful background information, particularly for work on cleaning, maintaining and adapting 'structures' as defined in the regulations (ie a wide range of buildings and infrastructure), and dealing with building services such as electricity, ventilation etc.

Under CDM 2015, risks to those using a structure as a workplace ('in-use' risks) need to be addressed by designers, taking account of the Workplace Regulations when carrying out their design. Exactly how this is done will vary depending on the situation, but it requires an understanding of the regulations.

Information about routine maintenance of items such as electrical or heating systems is normally provided by designers in the operation and maintenance (O&M) manual. The occupier will use this manual as an input into devising safe systems of work for these operations, taking account of other information in the H&S file.

Designer should, wherever possible, liaise with the future occupiers and their facilities managers, but it should be recognised that frequently designers will not have access to their detailed knowledge.

In all cases the primary responsibility for managing risk in the workplace will remain with the occupiers, but designers will have responsibility for ensuring that this can be achieved without undue difficulty in the structures that they have created.

Two examples are given here that demonstrate the importance of providing safe walking routes (access into buildings), where the combination of contamination (moisture) on footwear and the floor finishes may create a slip risk.

Example 1: Unsafe access

The entrance, featured in Lazarus *et al* (2006), has a small area of entrance matting adjacent to, and interspersed with, low slip-resistance tiling. During wet weather it can be unsafe to walk on.

Figure 1.1 Unsafe access into a building

Example 2: Safe access

This entrance has a large protective canopy and the walking surface within the entrance lobby is carpet. During wet weather it is safe to walk on.

Figure 1.2 Entrance to the QE2 Conference Centre, London

1.4 WHAT ARE 'WORKPLACES'?

Although CDM limits the 'in-use' consideration to structures 'designed as a workplace', with the possible exception of private domestic housing it is difficult to think of a structure that is not considered of necessity, on occasion, a workplace and designed as such. Even private domestic housing may not be excluded, as the responsibility of designers to consider risks to those carrying out future maintenance work etc has no limit under CDM (regardless of limits to other aspects of CDM). For example, a plumber working on a water-tank in a loft is then considered to be 'in a workplace', so it could be argued that consideration of workplace 'in-use' risks should apply to private housing too. Similar arguments can apply to streets, where there is cleaning activity or maintenance activity such as traffic light bulb replacement or crash barrier repair.

1.5 WHO ARE AFFECTED BY 'IN-USE' RISKS?

CDM 2015 refers to "any person using a structure designed as a workplace", which will include not only people who work there, but also others using the structure (often a building).

Although other people affected by the use of the structure (eg neighbours or passers-by) are not specifically mentioned, designers have a duty of care to them under the Health and Safety at Work Act 1974, and possibly other legislation (eg for environmental noise or chimney emissions). Many risks that affect the users of a building may also affect others (eg noise, gases, dust, air-borne pathogens).

Similarly, designers have a duty of care under the Health and Safety at Work Act 1974 to people using structures that are not designed as a workplace.

It is therefore necessary for designers to consider risks to all people, not just those "using a structure designed as a workplace" as required by CDM 2015.

1.6 WHAT ARE 'IN-USE' RISKS?

CDM 2015 does not limit the risks to be considered although it does state in guidance document L153 (HSE, 2015a) that the matters specifically requirements of the Workplace Regulations 1992 need to be complied with.

Workplace 'in-use' risks may involve a wider range of risks than those encountered in construction. However, the designer has to identify the risks and address them in a similar manner to construction risks, ensuring that their designs enable the occupiers to manage the residual risks. This guide provides information to assist that process and addresses a range of risks of both a specific and a general nature.

There are a great many generic and specific risks in the workplace, and they could be segmented in a variety of ways.

This guide identifies groups of types of risks and then addresses specific types of risk within that group. The groups are:

1. **Physical environment** (eg lighting, noise, vibration, temperature, wetness and humidity, draughts, confined spaces).
2. **Chemical and/or biological environment** (eg sanitary conditions, animals, moulds and fungal growths, hazardous materials and substances, smoke, dusts and fibres, other contamination/pollution).
3. **Hazardous systems** (eg electricity, hot water and steam, piped gases/liquids, hot surfaces, storage, radiation).
4. **Normal activities** (eg posture and manual handling, use of vehicles, use of plant and equipment, industrial processes, maintenance of highways, use of doors and windows/glazing, use of lifts escalators and moving walkways).
5. **Slips and trips** (eg while in motion on floors and ramps, or while using stairs or escalators).
6. **Working at height** (eg using access equipment such as ladders, at unprotected edges or adjacent to fragile surfaces).
7. **Abnormal events** (eg fire, explosion, falling objects, disproportionate collapse, drowning and asphyxiation, over-crowding, responding to emergency events, malicious human intervention).

Regardless of the range of workplace 'in-use' risks identified and examined in this guide, designers must consider the potential risks present in each specific situation. There may be others that are not examined here, but the process for considering them will be the same.

1.7 HOW ARE 'IN-USE' RISKS TO BE ADDRESSED BY DESIGNERS?

The core designer's duty is to identify risks, eliminate them if possible, reduce the level of residual risk so far as is reasonably possible (SFARP) and provide information for others. Designers need to consider risks that are foreseeable and significant in a proportionate manner as explained in L153. For further information see CIRIA C755 (Ove Arup, 2015).

Table 1.1 examines one particular 'in-use' risk. The table is intended to demonstrate the thought process and does not imply that a risk assessment has to be recorded in this manner or that the outcome would always be the same. See also Appendix A2 for a further example of the process.

Table 1.1 *Example of the consideration of a particular 'in-use' risk*

<table>
<tr><th colspan="3">Scenario: at a new high-street tyre shop, operatives will be required to load scrap tyres into 2.4 m high bins.</th></tr>
<tr><td>Risk identified</td><td colspan="2">Working at height and manual handling while disposing of scrap tyres. Manhandling tyres of various weights (increasingly heavy for SUVs) up to a platform and throwing them into the bin presents risk of muscular-skeletal damage and losing balance etc on the steps and platform.</td></tr>
<tr><td>Key features</td><td colspan="2">The risk of injury from a single operation is small, but operatives do this repeatedly and there has been, within the tyre shop group, examples of reportable accidents and time off work due to falls and sprains.</td></tr>
<tr><td rowspan="3">Risk management by the designers</td><td>Elimination of risk or substitution of lower-level risk; reduction of risk</td><td>The design team investigated the feasibility of sinking bins into a pit. This would cost £10k for the pit and its railing and drainage. There would also be a cost of £5k in adapting the skip delivery lorry so that it could load and unload at the pit. A time and motion study at an existing outlet showed that time saved would recoup the costs within five years, and there would be less risk of accidents and injuries leading to time off work. Operatives were supportive of the idea. Removal of the risk was the preferred and selected option.
The risk of working next to a pit was considered and safety railings were provided around the pit.</td></tr>
<tr><td>Provision of information</td><td>For management: requirement for annual inspection and if necessary maintenance of safety railings and pit drain.
For users: not required.</td></tr>
<tr><td>Control measures</td><td>The designer team suggested:
• provision of safe access into the pit for occasional clean-out
• provide a padlock to the access gate to prevent entry by unauthorised persons.</td></tr>
</table>

Note

Designers cannot be expected to set out how people are to behave or how they should be trained, instructed, provided with information and supervised, or what vehicles they should purchase etc. Such matters are for those responsible for operational matters or for the individual.

1.8 WHO ARE DESIGNERS?

The role of 'designer' is widely cast under the CDM regulations. Paragraph 72 of L153 (HSE, 2015a) makes it clear that this can include different people involved in the project, including a client.

In considering 'in-use' risks there will be many where the party acting as client has a central role to play (be it the end-user of a project or a project promoter who will be selling-on or renting in some manner). It is the role of the principal designer (PD) to ensure that this party (called the 'client' in this guide) makes appropriate inputs that are co-ordinated with the work of others.

In some situations, specialist inputs may be needed and, in particular, a facilities manager for the proposed building/structure should be involved in the design process if available. In certain circumstances it may also be appropriate to consult those who will work in a new or refurbished structure. The principal designer will need to ensure that the client's project manager takes the appropriate steps to involve these people.

> A developer client was constructing a factory, which was to be let to a company specialising in the zinc-coating of metal products. The special requirements of the process needed to be taken into account when selecting floor finishes and designing the extraction, alarm and guarding systems. The PD asked that the appropriate expertise be made available to the project and the client involved the future tenant for the necessary expertise.

1.9 PROVISION OF INFORMATION

Similar to construction, where the contractors will require information from the client and the designers, those responsible for managing and operating a completed structure will require information to enable them to discharge their duties under the Management of Health and Safety at Work Regulations 1999.

Information provided at the end of a construction project has traditionally been made available by the project team in the O&M manual. The H&S file is of a different nature, and must contain a range of

H&S-related information for those who will be involved in future cleaning, maintenance, refurbishment and demolition activities. The file may also refer to the O&M manual.

The manner in which the information is provided will need to be decided and take account of:

- the needs of the client, the occupier and the operator, where known
- who will hold the information and in what form
- how the information will be made available for reference
- arrangements for keeping the information updated.

Exactly how the H&S file is formatted will need to be agreed with the client, before documentation starts. The PD will be a key player in ensuring that appropriate decisions are made and carried out. A multidisciplinary approach will normally be necessary.

Further guidance about the H&S file is provided in Appendix 4 of L153 (HSE, 2015a).

An example of the type of information that the operator will find useful is provided in Table 1.2. It is not intended to be complete as each structure will present different requirements. However, it may assist in promoting thought.

Table 1.2 Examples of information arising from workplace issues

Topic	Information
Flooring (for each area/type, including stairs)	Slip Resistance Value (SRV) to be maintained. Maintenance regime to be followed
Fire	• compartmentation strategy and equipment • smoke management strategy and equipment • escape strategy, signage and equipment • eire-fighting strategy, information and equipment • emergency power system and fail safe provision.
Re-lamping	Strategy, requirements and means of access
Window maintenance and cleaning	Strategy, requirements and means of access
Cleaning facade	Strategy, requirements and means of access
Cleaning roof and gutters	Strategy, requirements and means of access
Cleaning floors, walls and ceilings	Strategy, requirements and means of access
Cleaning work surfaces	Strategy and requirements
Waste management	Strategy and provision
Process information	Summary of the process and information about the risks at the design stage
Ergonomics	Outline policy
Circulation strategy (for vehicles and pedestrians)	Strategy and provision
Storage and transportation strategy (for materials and products)	Strategy and provision
Separation from risks	Strategy and provision of controls including escape systems
Working at height	Strategy, requirements and means of access
Mechanical plant maintenance and replacement	Strategy, requirements and means of access
Façade maintenance and repair	Strategy, requirements and provisions built in
Inspection, maintenance and use of working platforms, including systems	Strategy, requirements and means of access
Roof access for maintenance and repair	Strategy, requirements and provisions built in

Topic	Information
Inspection, maintenance and use of restraint and suspension systems	Strategy, requirements and means of access
Asbestos management	Survey data, analysis results and condition records
Confined spaces	Identify spaces and residual risks, explain how access is: • prevented • permitted under controlled conditions, explaining the suggested precautions.
Crowds	Strategy and provision

Notes

1 The topics referred to include some arising from 'in-use' topics and some (cleaning and maintenance) that fall within the definition of 'construction' as defined by CDM 2015.

2 A topic would only apply when it is a significant feature of the workplace.

1.10 HOW TO USE THE 'IN-USE' GUIDANCE FOR DESIGNERS

This guidance should be used to prompt the thinking of the design team. It should not be used as a simple 'box-ticking' exercise. Each workplace will be unique, and it is the responsibility of the designers to respond to a particular situation in an appropriate manner.

One option will be for the design team to meet and, led by the PD, run through the risks that they have identified. The team could refer to this guide for further 'prompt' of other risks to be identified and, importantly, to start to assess the levels of risk and how the design will respond. This work should be carried out as an integral part of the design process and should be proportionate to the levels of risk involved.

It is important that consideration of the risks takes place and that appropriate decisions are made. It should be respected by designers and documented as appropriate for the project. This has already been happening with respect to cleaning and maintenance. Often, it has been through team discussion that sensible risk-reduction/management measures have been agreed and put into action. See Iddon and Carpenter (2009) for discussion of this aspect of the workplace.

Many risks will be seen to be adequately dealt with by simply complying with other regulations and by common-sense measures. Others may be more complex and, because of the severity of the risks involved, a more advanced approach may be required. As an example, one of the most common forms of accident is slipping or tripping over. As a result, the HSE sponsored the preparation of guidance by Lazarus *et al* (2006), which addresses the risks in detail and explains how they may be considered both at the design stage and subsequently 'in-use'.

As a further example, it is increasingly understood that long-term damage to health can arise through carrying out repetitive actions that are not unduly difficult or onerous in their own right. The construction industry has responded to this by measures such as adopting more mechanical lifting devices (eg for kerb stones) and reducing the weight of blocks to be handled and the normal weight of bags of cement and other products. Similar issues will arise in any workplace and designers should examine them during their consideration of 'in-use' risks, for example by reducing the requirement for manual handling.

Workplace 'in-use' issues that frequently arise include:

- selection of flooring materials so that they do not require high levels of grinding or buffing to maintain their performance over their design life
- sizing of lifts to avoid the need to manhandle objects going up stairs

- provision of permanently fixed access equipment to avoid having to manhandle moveable access equipment and siting of the permanent equipment so that it can be easily and safely accessed for maintenance, testing and use
- avoiding ladders and trap doors into plant rooms and onto roofs
- provision of lifting devices for raising tools and components to and from plant and equipment including in plant rooms
- provision of adequate working space in plant rooms and means of handling and installing replacement plant.

Similar issues will arise in most workplaces.

The guidance here may be used to assist risk identification and specific measures that need to be taken by the design team, as follows:

Chapter 2 summarises the specific requirements of the Workplace Regulations 1992 and the approved code of practice (ACOP) L24 (HSE, 2013a). While this chapter enables designers to rapidly gain an overview of the regulations, the team should be familiar with the requirements in detail.

Chapter 3 examines a range of risks and provides reference to further detailed guidance. The information provided should prompt discussion and further exploration of the issues. The team should include experts in all the areas relevant to the project, or additional input should be sought from appropriately experienced designers or experts.

Chapter 4 examines some typical workplaces, demonstrating the type of thinking required, which again the team should have expertise in for the type of workplace being designed.

Appendices A1 and A2 provide a checklist that may be used or adapted as a basis for recording consideration of risks, with an example.

Appendix A3 provides references that may need to be aware of, including requirements for particular activities and industries.

Note that disability is not specifically mentioned in the tables. The Building Regulations, the Workplace Regulations 1992 and the Equality Act 2010 all require these to be considered and this should be borne in mind when considering all risks.

Designers should use this guide to stimulate their consideration of risks based upon their own experience and expertise, and where they are unsure, expert advice should be sought. Note that the information given here is not intended to be prescriptive or to set in place requirements that designers have to follow, beyond the requirements of CDM 2015.

2 Summary of the specific requirements of the Workplace Regulations 1992 and its ACOP L24

In outline the Workplace Regulations 1992 apply to any workplace except:

- operational ships (covered by separate legislation)
- construction (covered by CDM 2015)
- mines (covered by separate legislation).

Certain workplaces are exempt from some of the requirements:

- Temporary worksites – Regulations 20 to 25 apply only so far as is reasonably practicable (SFARP).
- Workplaces in/on an operational aircraft, locomotive or rolling stock, trailer or semi-trailer – Regulation 13 applies and then only if stationary inside a workplace.
- Workplaces in the countryside (not inside a building) that are remote from an undertaking's main buildings – Regulations 20–22 apply and then only SFARP.
- Workplaces at a quarry or above ground at a mine – Regulation 12 applies and then only for a floor or traffic route inside a building.

Generally the requirements are absolute. Even where there is an exemption from this, the general risk management requirements of CDM 2015 will still apply. The requirements are very specific and, like the Building Regulations, designers need to be familiar with the requirements that impinge upon their areas of expertise. PDs need to be familiar with them all to enable them to manage interfaces between designers. The Workplace Regulations 1992 suggest that reference is made to local authorities and fire departments, as well as to the wealth of guidance published by the HSE (see Websites).

For the purposes of this guide, a summary of the requirements expressed in the Workplace Regulations 1992 and the associated ACOP L24 (HSE, 2013a) is given in Table 2.1. The information provided paraphrases the requirements to give an easily-accessed view. It can be seen that the requirements reflect the standard of provision that is expected in today's society where everyone's needs should be respected and met in a sensible manner. They align with the Building Regulations, albeit with some additional requirements and advice reflecting workplace requirements. They are directed towards people at work. In many workplaces there will be others present for a variety of reasons (students, patients, visitors etc as well as passers-by), and their requirements will also need to be met in a sensible manner. Similarly, although mobility-impaired persons are specifically mentioned in some places, design in its generality should take account of the needs of all people.

Also note that 'sick building syndrome' (SBS) is not mentioned. This is believed to be because SBS is not currently considered a recognised illness. For more information see HSE (1995a).

Table 2.1 Summary of the Workplace Regulations 1992

Reg	Summary of requirements	Notes
1–4		These are opening sections and do not contain specific technical requirements.
4a	***Safety and solidity*** A building used as a workplace must have appropriate 'safety and solidity'.	Compliance with building regulations and design codes will normally be adequate, but consideration should be given to any particular hazards identified.
5	***Maintenance*** A workplace and its equipment and systems must be properly cleaned and maintained.	Designers need to consider how the cleaning and maintenance of the structure (including its systems), can be done with minimum risk and what information needs to be provided, what needs to be done, and how to do it safely.
6	***Ventilation*** An enclosed workplace must be adequately ventilated with fresh (uncontaminated) air without excessive drafts, smells or humidity and without risk of Legionnaire's disease.	For H&S purposes, alarm systems may be necessary. There are some exceptions to the requirement for ventilation, where the work is in a 'confined space' and special measures are taken such as the provision of breathing apparatus, although such spaces, or the need for staff to enter them, should be designed out where reasonably practicable.
7	***Temperature*** A workplace must be at a reasonable temperature without problems from escaping vapours emanating from the heating or cooling system, and with thermometers to keep a check.	Thermometers will normally be part of the fit-out.
8	***Lighting*** A workplace must be adequately lit, with natural lighting SFARP, and with suitable emergency lighting where loss of lighting would present a risk.	Work-spaces without good natural lighting should normally be avoided. Emergency lighting beyond what is required for safe egress in a fire will be dictated by the work processes.
9	***Cleanliness and waste*** The structure and its contents need to be kept clean and waste dealt with.	Designers need to consider how the surfaces of the structure can be cleaned with minimum risk. Information on what needs to be done and how to do it safely should be provided. Designers also need to consider how waste will be managed in a safe manner. Information on the waste management strategy and what needs to be done should be provided.
10	***Room size*** People need 'sufficient room' while they work.	The basic requirements should be exceeded, particularly if space is taken up by plant and equipment.
11	***Workstations and seating*** People need to be able to work in 'adequate comfort' (including protection from the weather), and able to work without strain.	Mainly operational in intent – but designers who become involved in designing workstations and work processes need to have expertise in ergonomics and particularly manual handling and activities that are repetitive or awkward or involve vibration, or seek help.
12	***Floors and traffic routes*** Workplace floors and walking or vehicle routes need to be 'safe and fit for purpose'.	Focused mainly upon floors having an adequately smooth and low-slip-risk finish and being adequately drained, with handrails to stairs and ramps, ramps not being too steep and holes being guarded. Walking or vehicle routes to be designed to be clear of obstruction, especially in busy areas/escape routes.

Reg	Summary of requirements	Notes
13	***Falls or falling objects*** People need to be protected from falls or falling into dangerous places or material falling onto them.	Covers a range of risks and the need for: • fencing and toe-boards at edges of floors • fencing or covering of hazardous materials (including those in tanks, pits, vessels etc and including protection from vapours) • fencing to protect from vehicles • where possible, protection even where there is normally no access • special protection at loading bays etc to protect from falls while enabling vehicle access, eg by a mechanised system • where possible, staircases rather than ladders • fixed ladders to have max 6 m between landings, with safety hoops and gated access • no fragile materials and roofs or other covers and all such areas to be accessed easily • no changes in level that could cause a fall • safe storage and stacking of materials/products • safe work practices that do not require people to clamber over (eg flat-bed lorries) and loads, ie where they may fall. • Some rules of thumb are set out.
14	***Glass (windows etc)*** People need to be protected from 'breakage' and 'recognise the risks presented'.	Each situation needs to be risk assessed and suitable materials selected and protection provided. Some rules of thumb are set out. Manifestation may be considered.
15	***Opening windows etc*** People need to be protected from 'falling out of openings and colliding with windows etc' when open.	Each situation needs to be risk-assessed both for 'in-use' conditions and during operation of the open/close mechanism.
16	***Cleaning windows etc*** People need to be able to clean windows etc safely.	Each situation needs to be risk-assessed, avoiding the need for cleaning (eg self-cleaning glass), or providing the ability to (SFARP) clean from the inside or (failing that) from the outside using suitable equipment (cradles or travelling ladders with attachment for safety harness) and/or fixed safe anchorage points for harnesses or long ladders (6 m to 9 m max). (The use of external mobile equipment is not mentioned).

Reg	Summary of requirements	Notes
17	***Organisation of traffic routes*** People and vehicles need to be able to circulate safely. Traffic routes include stairs, ladders, ramps etc, and loading bays together with associated doors and gates.	The Regulations require adequate provision as well as organisation. The ACOP contains considerable detail on what is a specialist subject. Matters to consider include: • adequate provision and organisation of routes • adequate separation from adjacent work activities • adequate width at doors and gates • separation of pedestrians and vehicles • signing, speed limits, humps etc – various issues • avoidance of obstructions, low head-room etc • provision for the needs and requirements of those with impairments • avoid ladders/steep stairs if possible and especially if they cannot be used safely (including the need to carry loads) • restrict vehicle types to areas where they can operate safely • allow for two-way traffic plus parked vehicles • provide protection from vehicle impact as necessary and protect vehicles at edges, drops etc • protect people from fumes or risk from shed loads • avoid the need to reverse (or take precautions) • avoid trapping risks – especially with driverless vehicles • physically separate/protect people from vehicles at doorways, tunnels, bridges etc • consider risks where vehicles are loaded/unloaded/tip etc • consider risks at crossing points and seek grade separation (see particular requirements on this) • consider crowds at start/end of day/shift • consider risks in loading bays, also consider refuges.
18	***Doors and gates*** They must be safe in use, ie suitably constructed with adequate safety devices as part of the finished installation.	The Regulations and ACOP point out a range of issues: • if running on tracks, have a device to stop them coming off • if vertically opening, have a device to prevent it falling back down • if powered: • have a device to prevent trapping of people • have an emergency cut-out switch • be capable of being opened manually (but safely in the event that power is restored) unless it opens automatically if the power fails. This requirement excludes doors/gates that are there to protect from falls, eg lift doors • if it could be pushed open and hit people, have a vision panel (which enables wheelchairs to be seen) • if tools are required for manually operation, they should be available at all times Detailed advice is given.
19	***Escalators and moving walkways*** They must function safely.	The Regulations note that they must have "any necessary safety devices" and "be fitted with one or more emergency stop controls which are easily identified and readily accessible".
20	***'Sanitary conveniences', ie toilets*** They are to be 'suitable, sufficient, readily accessible'.	They must be in adequately-lit and ventilated rooms and provide separate male/female facilities or provide a single-user room that can be secured from the inside. Ensure adequate privacy and also appropriate separation from areas where food is processed, prepared or eaten. Each WC must be in a single-user room, which can be secured from the inside. The minimum extent of provision is given in Reg 21, which also sets out the standard of provision (ie drained, flushable, properly-equipped, ventilated, lit and protected from the weather, odours dealt with).

Reg	Summary of requirements	Notes
21	***Washing facilities*** In addition to being provided at toilets and changing rooms, suitable and sufficient readily accessible washing facilities, including showers if necessary, are to be provided.	Appropriate washing facilities (including baths or showers if required by the nature of the work), must be provided at sanitary conveniences and also near changing rooms. Ensure adequate privacy. Basic requirements for provision are: • provide separate male/female facilities or provide a single-user room that can be secured from the inside • sufficient for all to use without undue delay • special provision for mobility-impaired people • shot and cold running water with anti-scald mixer • soap or other means of getting clean • stowels or other means of drying • ventilated, lit and protected from the weather. Minimum numbers of facilities are set out in detail and guidance given for remote workplaces and temporary worksites. Warning is given about Legionnaire's disease.
22	***Drinking water*** People must be provided with 'wholesome drinking water' that is readily-accessible at suitable places and marked as fit for drinking.	Consider risks of contamination from chemicals, bacteria etc.
23	***Accommodation for clothing*** Provide suitable and sufficient storage for clothes.	Consider the need to accommodate both clothing removed on arrival and work clothes removed before departure – separately if there is a contamination risk. Provide storage that enables people to hang clothing in a clean, warm, well- ventilated dry place where it can dry out. Consider the need for secure storage and additional provisions for PPE.
24	***Facilities for changing clothing*** Provide changing facilities if people need to change into special clothing for work and need privacy to do so.	Provide separate facilities for men and women. Consider the risks of cross-contamination. Careful consideration needs to be given to the relationship between changing rooms and other facilities (clothing storage, toilets, workrooms, eating facilities etc). The extent of provision will require a study of possible numbers and work patterns. No specific mention is made of changing (and other facilities) for cyclists, but this could be considered.
25	***Facilities for rest and to eat meals*** Provide "suitable and sufficient rest facilities at readily accessible places".	The detailed requirements must be carefully studied, taking account of working conditions, food contamination risks, the particular requirements of pregnant women and nursing mothers and the need to avoid nuisance from tobacco smoke.
25a	***Disabled persons*** The needs of disable persons must be catered for.	This involves access, welfare facilities and workstations in areas of a workplace that they use or occupy.

Note

Regulations 4a and 25a were introduced by amendment in 2002.

3 Information about risks

The following tables address a range of risks and provide information to assist understanding and thinking. It is not (and never could be) comprehensive, but should assist in informing designers and helping them to access guidance. Space is provided for users to add their own notes and references.

The control measures shown are typical controls that may be chosen and implemented by those designing and/or managing a workplace.

References and websites are provided in a separate section.

Physical environment – lighting (natural and artificial)		
Risk		Insufficient light for safe activity that can cause headaches or damage to sight due to prolonged periods in dark spaces or low quality light source. Other problems are glare and flashes.
Key features		Levels of lighting, quality of the light, steadiness of the light.
Triggers		Inadequate intensity or quality of light. Excessive light.
Background		Adequacy of light is normally assessed using the information provided in CIBSE guidance. The medical requirement for daylight appears to be uncertain.
E	*Elimination*	Inappropriate light conditions can easily be removed as a risk by making proper provision and/or protection. Incidences of unshielded glare or flash should be eliminated SFARP.
R	*Reduction*	Minimise exposure high/low levels of light.
I	*Information for users*	Warning notices for glare and flash
	Information for others	Basis of design statement in the O&M manual. Maintenance requirements set out in the O&M manual.
C	*Controls envisaged*	Shading and eye protectors for glare and flash.
Key references		• CIBSE (1983, 1996, 2012a, 2012b) • HSE (1997a)
Notes		

Physical environment – noise		
Risk		Impairment of hearing due to prolonged periods at excessive noise levels or short-term exposure to extreme levels of noise. Lack of audibility. Apart from damage to hearing, noise may contribute to stress-related ill-health and noise may cause annoyance and distraction from other risks.
Key features		Noise is a common problem with different effects at different frequencies and cumulative over time. It is a specialist subject. The effect of noise sources in the workplace is influenced by the degree of attenuation and transmission between spaces. The required limits on noise will depend on the situation.
Triggers		Machinery, traffic, processes.
Background		Limits on levels of noise in workplaces are stated in the Control of Noise at Work Regulations 2005, with action values and limiting values. Building Regulations Part E (HM Government, 2003) deals with acoustic insulation between domestic spaces and reverberation within certain common spaces and classrooms. These are normally issues of comfort rather than H&S, although in some circumstances they could cause distraction. Noise may affect adjacent buildings – in which the tolerance of noise may be low (eg hospitals). Requirements for comfort are given in Hawkins (2011)
E	*Elimination*	Reduction in noises generated to a safe and acceptable level or separation of people from excessive levels of noise are the best design options.
R	*Reduction*	The level of risk from residual noise may be reduced by applying appropriate controls.
I	*Information for users*	Advisory notices.
	Information for others	Noise nuisance management strategy.
C	*Controls envisaged*	Attenuation measures and barriers. Limiting periods of exposure. Noise monitoring devices that alert people to a rise in noise levels. PPE/noise protectors to protect people's hearing, preferably to be used for short periods only for specific tasks in noisy spaces.
Key references		• CIBSE (2004) • Hawkins (2011) • HM Government (2003) • HSE (1998a, 2012a) ***Statutes*** • Control of Noise at Work Regulations 2005 ***Websites*** • HSE *Noise at work*: **www.hse.gov.uk/noise** • Institute of Acoustics: **www.ioa.org.uk**
Notes		

Physical environment – vibration		
Risk		Discomfort. Whole-body vibration (WBV). Hand-arm vibration syndrome (HAVS) and 'white finger'. Distraction from various other risks. Damage to fabric, eg fixings coming loose and causing accidents.
Key features		Vibration is a risk in certain specific circumstances, mainly industrial. It is a specialist subject. Limits on levels of vibration in workplaces are stated in the Control of Vibration at Work Regulations 2005, with action values and limiting values. Vibrations may affect adjacent buildings – in which the tolerance of vibration may be low (eg hospitals).
Triggers		Machinery, tools, traffic, processes.
Background		Safe levels of vibration are normally assessed.
E	*Elimination*	Reduction in vibrations generated to a safe level or separation of people from excessive levels of vibration are the best design options. Remedies include the provision of anti-vibration mounts and dampers.
R	*Reduction*	The level of risk from residual vibration may be reduced by applying appropriate controls.
I	*Information for users*	Advisory notices.
	Information for others	Vibration management strategy.
C	*Controls envisaged*	Attenuation measures, eg anti-vibration mounts, dampers, isolation barriers. Vibration monitoring devices that alert people to a rise in levels of vibration. Limited periods of exposure.
Key references		• HSE (2005a, 2012b) • Scarlett and Stayner (2005) ***Statutes*** • Control of Vibration at Work Regulations 2005 ***Websites*** • HSE *Vibration at work*: **www.hse.gov.uk/vibration**
Notes		

Physical environment – temperature		
Risk		Physical symptoms and possibly heat-stroke/frost-bite or other severe medical conditions. Worker fatigue/loss of concentration that may cause mistakes leading to accidents. Loss of dexterity that may cause accidents with controls or sharps or hazardous materials etc. Distraction from various other risks.
Key features		Thermal comfort is a normal part of building services design. Hot processes, cold stores etc need to be carefully considered, including the effects of any PPE worn by workers.
Triggers		Extreme hot or cold (or less extreme conditions but for prolonged periods). Minimum of 16°C for occupied areas in HSE (2013s) L24. A maximum of 30°C has been suggested for offices.
Background		Normally the temperature range experienced by people is controlled within acceptable levels by heating, ventilation and comfort cooling or air conditioning. Note that ventilation also serves to remove stale air containing carbon dioxide and other contaminating gases. There are no absolute requirements in the Building Regulations or Workplace Regulations 1992 for a maximum temperature, but Hawkins (2011) sets out suggested limits for design. In safety critical situations where a loss of concentration could affect safety risk, the provision of cooling may be appropriate.
E	*Elimination*	Provision of adequate insulation and heating/cooling systems.
R	*Reduction*	The level of risk from residual risks may be reduced by applying appropriate controls.
I	*Information for users*	Advisory notices and thermometers.
	Information for others	Normally provided within the O&M manual.
C	*Controls envisaged*	Provision of PPE to keep warm may be appropriate in some circumstances. In extreme conditions (eg cold-stores, furnaces), there could be risk of death and appropriate safeguards (including limited periods of exposure) and alarms need to be designed in. Provision of PPE to keep cool is rare, although local area cooling or PPE may be provided.
Key references		• BSRIA (1988) • CIBSE (2004, 2015) • HM Government (2010a) • HSE (2000a, 2013a, 2013b) ***Statutes*** • BS 7915:1998 • Workplace (Health, Safety and Welfare) Regulations 1992 ***Websites*** • HSE *Heat stress*: **www.hse.gov.uk/temperature/heatstress**
Notes		

Physical environment – wetness and humidity		
Risk		Presence of moisture as water or dampness or moisture vapour OR absence of moisture.
Key features		Can cause medical problems and may exacerbate existing conditions. Moisture can make floors slippery. Dryness can cause drying of the mucous membranes.
Triggers		Exposure to moisture/dryness.
Background		In most structures exposure to moisture as water or dampness is designed out. Levels of moisture vapour are normally controlled by ventilation or air conditioning. Moisture drips from condensation or escaping from air conditioners may also cause slippery floors. The Building Regulations Parts C and F include provisions to exclude entry of moisture and water and to manage it by ventilation.
E	*Elimination*	Appropriate insulation/membranes and heating/air conditioning systems.
R	*Reduction*	The level of risk from residual risks may be reduced by applying appropriate controls.
I	*Information for users*	Not normally required.
	Information for others	May be dealt with in the O&M manual.
C	*Controls envisaged*	Provision of shields or PPE to keep dry may be appropriate in some circumstances.
Key references		• CIBSE (2004) • HM Government (2004a and 2010a) • HSE (2000a)
Notes		

Physical environment – draughts		
Risk		Noticeable movements of air through occupied spaces.
Key features		Can cause medical problems.
Triggers		High airflows due to ventilation (including air conditioning) or at openings or local to air-leaks.
Background		The avoidance of draughts is normally achieved through seeking to reduce energy consumption for heating.
E	*Elimination*	In most structures exposure to draughts is designed out.
R	*Reduction*	The risk is normally designed out, but if it remains then its effect may be reduced by design of enclosed work spaces (eg cubicles) or screening.
I	*Information for users*	Not normally required.
	Information for others	Not normally required.
C	*Controls envisaged*	Provision of special clothing may be appropriate in certain spaces.
Key references		• HSE (2000a) • HM Government (2010a)
Notes		

Physical environment – confined spaces		
Risk		Risks include oxygen deprivation, toxic atmospheres, flammable or explosive atmospheres, oxygen enrichment, hostile environments, restricted access and egress, infectious disease, hazardous residues, weather, noise.
Key features		A confined space is any place where there is a foreseeable risk due to lack of a free flowing supply of breathable air, dangerous gases, vapours, fumes, residues etc, engulfment by liquids or unstable granular materials, restricted access or egress, plant or machinery close by, heat stress. A confined space does not have to be small or fully enclosed and can be above or below ground.
Triggers		Inadequate ventilation or failure of ventilation system. Activities or other disturbance or use of equipment that causes changes in the atmosphere. Use of unsuitable equipment in hazardous atmospheres.
Background		Examples of confined spaces include chambers and inspection pits, manholes, sewers and drains, pipes, tunnels, cellars, tanks, silos, process vessels, boilers, underground boiler houses, some plant and motor rooms, flues, ducts, box girder bridges etc.
E	*Elimination*	Ensure adequate ventilation of plant rooms, switch rooms, basements etc to avoid creating confined spaces. Design to allow for the withdrawal of plant and equipment from confined spaces for maintenance and cleaning. Design for remote monitoring of equipment in confined spaces. Avoid specifying equipment that is likely to create a hazardous atmosphere within a confined space.
R	*Reduction*	Design to minimise maintenance and cleaning requirements on plant and equipment in confined spaces. Design in suitable isolation measures for plant and equipment in confined spaces. Design in alarm/detector systems or other safety devices where appropriate. Design in sufficient manholes for ventilating tunnels, pipes etc likely to require access. Design in suitable and sufficient means of access/egress for confined spaces, including for use in emergency escape and rescue scenarios. Ensure adequate working space to carry out maintenance, cleaning and other relevant activities.
I	*Information for users*	Specify appropriate signage to highlight confined spaces.
	Information for others	Clearly identify confined spaces and any measures designed in to minimise risks, ie on drawings and in the H&S file.
C	*Controls envisaged*	Compliance with the Confined Spaces Regulations 1997 and HSE (2014a) L101.
Key references		• HSE (2014a) ***Statutes*** • Confined Spaces Regulations 1997
Notes		

Chemical and/or biological environment – sanitary conditions		
Risk		Spread of human bacterial contamination and other sanitary issues such as control of vermin.
Key features		Can cause medical problems.
Triggers		Lack of adequate sanitary facilities and inadequate waste storage facilities.
Background		Provision of adequate toilets, washing facilities and, where necessary, provision for changing of clothing is required by Building Regulations Part G (HM Government, 2010b) and the Workplace Regulations 1992. The Building Regulations also require that solid waste materials can be stored properly.
E	*Elimination*	Proper facilities should always be designed in or otherwise provided.
R	*Reduction*	The level of risk is normally reduced SFARP.
I	*Information for users*	Advisory notices.
	Information for others	In the O&M manual if required.
C	*Controls envisaged*	In high-risk situations (eg hospitals) measures such as barrier clothing and hand-cleansing gels may be required.
Key references		• ACDP (2003, 2005) • CIBSE (2014) • DOH (2013) • HM Government (2010b) • HSE (2013a) ***Statutes*** • BS 6465-1:2006+A1:2009 • Workplace (Health, Safety and Welfare) Regulations 1992
Notes		

Chemical and/or biological environment – animals		
Risk		Presence of animals in spaces occupied or visited by people. Residual risks from dead animals or animal-based products or residues.
Key features		Can cause medical problems. Risk of contamination, eg transfer of parasites, allergic reactions, bronchial problems, viral infections etc.
Triggers		Presence of insects, vermin, viruses, spores, excrement, bird droppings etc.
Background		The presence of vermin is normally designed out, although some places are particularly prone to infestation (eg food stores, kitchens, eating areas and closed-off spaces such as lofts). The presence of insects can be designed out to some extent. The presence of other animals (by choice) requires specialist knowledge and management. The presence of pre-existent triggers such as anthrax spores should be investigated by experts where the risk is identified.
E	*Elimination*	The ingress and support of vermin should be designed out, but complete success cannot be guaranteed. Avoid hidden voids that cannot be cleaned out. Avoid materials that can harbour infestation. When working with old materials, test for pre-existent risks and eliminate them where appropriate.
R	*Reduction*	Further risk reduction is an 'in-use' management issue.
I	*Information for users*	Advisory notices.
	Information for others	Particular, unusual issues may be dealt with in the O&M manual.
C	*Controls envisaged*	Design to facilitate inspection and cleaning in high-risk areas.
Key references		• HSE (1997b, 2001a, 2012c) • Newton *et al* (2011)
Notes		

Chemical and/or biological environment – moulds and fungal growths		
Risk		Presence of moulds and/or fungal growths in spaces occupied or visited by people.
Key features		Can cause medical problems. Risk of allergic reaction.
Triggers		Presence of spores, growing medium and supportive environmental conditions. The issue of Aspergillus spores, which may be released during refurbishment work, has been identified as a particular risk to the elderly and infirm.
Background		The presence of fungal growths is normally designed out, although some places are particularly prone to outbreaks (eg damp spaces and closed-off spaces such as lofts).
E	*Elimination*	The ingress and support of fungal growths should be designed out, but complete success cannot be guaranteed. Avoid hidden voids that cannot be cleaned out. Avoid materials that can harbour infestation.
R	*Reduction*	Further risk reduction is an 'in-use' management issue.
I	*Information for users*	Advisory notices.
	Information for others	Particular, unusual issues may be dealt with within the O&M manual.
C	*Controls envisaged*	Design to facilitate inspection and cleaning in high-risk areas.
Key references		• BRE (1992) ***Websites*** • Fungal Infection Trust: **www.fungalinfectiontrust.org** • HSE *Using biocides*: **www.hse.gov.uk/biocides/using.htm** • National Aspergillosis Centre: **www.nationalaspergillosiscentre.org.uk** • RICS: **www.rics.org/uk**
Notes		

Chemical and/or biological environment – other hazardous materials (solids, liquids, gases, fumes)		
Risk		Contact with hazardous materials. Spread of escaped liquids and gases. Gases given off during activities such as welding or soldering. Exhaust gases and gases from malfunctioning boilers etc mainly carbon dioxide.
Key features		Can cause medical problems if liquids or gases escape or are released and are ingested orally or through the skin. May cause allergic reaction. Can cause explosions.
Triggers		All chemicals and other materials of a hazardous nature whether imported or created. Processes or activities that use or release gases or liquids.
Background		All hazardous materials need to be considered, in particular where there are industrial processes or experimental work. Gases such as radon or methane emanating from the ground need to be considered. The Building Regulations Part C deal with this. For radon, see also the section on radiation. The build-up of gases such as carbon dioxide (from breathing) and of the products of combustion (eg carbon monoxide, which can kill) and fuel storage need to be dealt with by proper ventilation and the Building Regulations Parts F and J deal with this. Gases emanating from the structure itself should not be a problem if acceptable materials are used. Some solvents have been suspected of causing problems. The Building Regulations Part C deals with the risk from gases from the ground and Part D deals specifically with risks from formaldehyde foam insulation.
E	*Elimination*	Where possible hazardous materials should be removed from areas accessed by people. SFARP avoid specifying materials/substances known to be hazardous. Specification of materials that can only be cleaned using hazardous processes or compounds should be avoided.
R	*Reduction*	Substitution of a lower-risk may be possible. The overall final level of risk is normally reduced SFARP.
I	*Information for users*	Alarms. Advisory notices and test equipment.
	Information for others	Normally provided within the O&M manual.
C	*Controls envisaged*	In dangerous situations (eg nuclear) periodic testing may be appropriate and PPE may be needed.
Key references		• CDC (2005) • CIBSE GS A4 • HSE (2000a, 2000b, 2006b, 2011a, 2011b, 2013c, 2013d, 2013e) • HM Government (1992, 2003, 2004a, 2010c) • HSE (2011b) • TRADA (2005) • HSE (2006b) ***Statutes*** • Control of Asbestos Regulations 2006 • Control of Lead at Work Regulations 2002 • COSHH Regulations 2002 • CHIPS Regulations 2009 • Gas Safety (Installation and Use) Regulations 1998 • Ionising Radiations Regulations 1999 ***Websites*** • HSE COSHH basics: **www.hse.gov.uk/coshh/basics.htm** • HSE COSHH essentials for welding, hot work and allied processes: **www.hse.gov.uk/pubns/guidance/wlseries.htm** • US National Library of Medicine TOXNET Toxicology Data Network: **http://toxnet.nlm.nih.gov** Note: the HSE has published a wide range of guidance notes on this subject on its website.
Notes		

Chemical and/or biological environment – pollution (smoke)		
Risk		Creation and spread of smoke (ie air-borne particles and gases).
Key features		Smoke created deliberately or accidentally can cause medical problems if gases and particles escape or are released and are ingested orally or through the skin. May cause allergic reaction. Can cause blindness and incapacity in an emergency situation.
Triggers		Processes or activities which release smoke (as well as fire).
Background		The presence of smoke is normally considered in particular industrial processes or experimental work – and in smoking, which is now recognised as dangerous.
E	*Elimination*	Where possible smoke should be removed from areas accessed by people. Dusts should be removed before emission.
R	*Reduction*	For process situations substitution of a lower-risk may be possible. For accidental situations, use of materials that are less combustible or give off less dense smoke or less dangerous gases and particles may be appropriate. The level of risk is normally reduced SFARP.
I	*Information for users*	Advisory notices and test equipment.
	Information for others	Normally provided within the O&M manual.
C	*Controls envisaged*	In dangerous process situations periodic testing of the air may be appropriate and PPE may be needed. Control of accidental fire situations – see also the section on fire. Risks to environment to be managed by analysis of chimney height and flume behaviour.
Key references		• HM Government (2006, 2010a) *Statutes* • BS EN 12101-6:2005
Notes		

Chemical and/or biological environment – pollution (dusts and fibres)		
Risk		Spread of dusts and fibres.
Key features		Can cause medical problems if dust particles become airborne or contaminate clothing and are ingested orally or through the skin. May cause allergic reaction or cancer. Dust/air mixtures may be explosive – see the seciton on explosion.
Triggers		Processes or activities that use or release dusts or fibres. Note: if in doubt, research the subject, including agricultural products, wood products, cement, stone, silica etc.
Background		The presence of dusts or fibres is normally considered in particular industrial processes or experimental work. However, asbestos must always be borne in mind and dealt with in accordance with the latest regulations and guidance.
E	*Elimination*	Where possible dangerous dust should be removed from areas accessed by people. Materials likely to create hazardous dust should be avoided SFARP. If pre-existing, eg asbestos in refurbishment projects, careful consideration should be given to their controlled removal.
R	*Reduction*	The level of risk is normally reduced SFARP although in the case of specific materials such as lead, asbestos and nuclear materials, legal requirements must be met.
I	*Information for users*	Advisory notices and test equipment.
	Information for others	Normally provided within the O&M manual.
C	*Controls envisaged*	In dangerous situations (eg nuclear) periodic testing of dust may be appropriate and PPE may be needed. Periods of exposure may be limited. Oxygen monitors may be provided.
Key references		• HSE (2000b, 2002b, 2002a, 2011a, 2013c, 2013d, 2013f, 2013o) ***Statutes*** • COSHH Regulations 2002 • Control of Asbestos Regulations 2006 • Control of Lead at Work Regulations 2002 • Ionising Radiations Regulations 1999 ***Websites*** • HSE *Dust*: **www.hse.gov.uk/construction/faq-dust.htm**
Notes		

Chemical and/or biological environment – other contamination/pollution		
Risk		Chemical or bacterial contamination.
Key features		Can cause medical problems if fine particles are spread and contaminate clothing etc, and are ingested orally or through the skin. May cause allergic reaction.
Triggers		Presence of contamination in air, soils, groundwater, raw materials. Materials, processes or activities that release particles or sprays. Lack of safe processes or use of inappropriate surfaces.
Background		Testing for contamination is required wherever it could present a significant risk. Safe methods for using surfaces have been developed in the food industry and in specialist areas such as laboratories. Control of pollution from the ground is dealt with in the Building Regulations Part C (HM Government, 2004). Bacterial contamination includes Legionellosis. This is a specialist subject.
E	*Elimination*	It may be possible to remove a potential source of contamination.
R	*Reduction*	Substitution of a lower-risk may be possible. The level of risk is normally reduced SFARP taking account of statutory and advisory limits.
I	*Information for users*	For residual risks, advisory notices and test equipment.
	Information for others	Normally provided within the O&M manual.
C	*Controls envisaged*	In dangerous situations, periodic testing may be appropriate. Systems may be made safer to reduce risks of spillage, particularly during maintenance, eg by providing for isolation of runs. PPE may be needed.
Key references		• CIBSE (2013) • HM Government (2004) • HSE (1997c, 2002a, 2006c, HSE (2009a), 2013c, 2013g) • Wilson *et al* (2007) ***Statutes*** • Control of Lead at Work Regulations 2002 • COSHH Regulations 2002 • Industry-related regulations may apply, eg catering, laboratories, farming ***Websites*** • Gov.uk *Contaminated land*: **https://www.gov.uk/contaminated-land/overview**

Notes

Hazardous systems – electricity		
Risk		Electrical shock.
Key features		Causing death or injury.
Triggers		Contact with live wiring or equipment. Discharge of static electricity.
Background		In the UK, the requirements of BS 7671:2008+A3:2015 have to be complied with and the Building Regulations Part P (HM Government, 2013a) apply to dwellings.
E	*Elimination*	Removal of electrical systems is not feasible, but in high-risk areas (wet, explosive gases or dusts) special measures should be taken. Measures to protect against electrocution and allow adequate isolation should be designed in SFARP, eg special fittings, double insulation, locked cabinets, interlocks and isolation systems.
R	*Reduction*	Low voltage electricity systems.
I	*Information for users*	Advisory notices.
	Information for others	An essential part of the O&M manual.
C	*Controls envisaged*	Special fittings, double insulation, locked cabinets, interlocks and isolation systems etc.
Key references		• CIBSE (1986) • HM Government (2013a) • HSE (1989, 2013h) • Scaddan (2011) *Statutes* • BS 7671:2008+A3:2015 • BS 5958-1:1991 • Electricity at Work Regulations 1989
Notes		

Hazardous systems – hot water and steam (and steam condensate)		
Risk		Excessively hot water (above 43°C) or steam. Steam condensate may be at high temperature (and pressure).
Key features		Can cause severe burns, scalds and death. Steam burns twice, on contact and on condensing, causing deep burns. In industrial situations, steam is usually under pressure and temperatures can be over 100°C. Hot water assists legionella growth.
Triggers		Contact with excessively hot water or steam.
Background		Limit temperatures of free water and steam are normally set so as not to cause problems unless there is an accidental escape.
E	*Elimination*	Limit hot water temperatures SFARP.
R	*Reduction*	Protection against accidental escape should be to a high level of reliability. Reliance upon controls is not recommended.
I	*Information for users*	Warning signs where appropriate.
	Information for others	Instructions for maintenance and operation.
C	*Controls envisaged*	Protective shields, PPE.
Key references		• BP Safety Group (2004)
Notes		

Hazardous systems – piped gases/liquids		
Risk		Dangerous gases and liquids. Vacuum/suction lines and systems.
Key features		Injury, medical conditions or death.
Triggers		Escape of dangerous gases/liquids. Blasting with carried materials, eg grit.
Background		Expertise in the handling of gases/liquids resides within specialist companies. Liquids may add to electrical risks, eg drains running through switch rooms.
E	*Elimination*	Not normally possible.
R	*Reduction*	Protection against accidental escape should be to a high level of reliability.
I	*Information for users*	Routing drawings, physical identification and isolation points. Warning signs where appropriate.
	Information for others	Instructions for maintenance and operation.
C	*Controls envisaged*	Provision of masks and breathing apparatus in extreme situations. Eye wash, safety showers. Provision of alarms. Physical shrouding to protect from damage (particularly when exposed, but also for protection of pipes carrying dangerous gases, even within walls and pep ways).
Key references		• IChemE (2004–2012) • EI (1981, 1991, 1998) • Mather and Lines (1999) • HSE (1997d)
Notes		

Hazardous systems – hot/cold surfaces		
Risk		Excessively hot surfaces (above 43°C). Excessively cold surfaces.
Key features		Can cause severe burns (including cryogenic burns) and death. The young, elderly and infirm are at particular risk.
Triggers		Contact with excessively hot or cold surfaces.
Background		Limit temperatures of surfaces accessible to the touch are normally set so as not to cause problems unless there is an accidental escape.
E	***Elimination***	Surfaces that do not need to be exposed should be covered/insulated/lagged. Limit hot water temperatures SFARP.
R	***Reduction***	Protection against accidental touching should be to a high level of reliability.
I	***Information for users***	Warning signs where appropriate.
	Information for others	Instructions for maintenance and operation.
C	***Controls envisaged***	Provision of PPE may be appropriate.
Key references		***Websites*** • RoSPA *Home safety*: **www.rospa.com/home-safety**
Notes		

<table>
<tr><th colspan="3">Hazardous systems – storage</th></tr>
<tr><td colspan="2">Risk</td><td>Storage of materials presents three main areas of risk, which are the strength and stability of the storage system, the risks involved in the loading and unloading of materials into and out of storage and the actual materials.</td></tr>
<tr><td colspan="2">Key features</td><td>Storage systems involve a wide range of situations, including the storage and maintenance (often at height) of materials and products that may be hazardous, heavy or unwieldy and may contain stored energy that may be accidentally released.</td></tr>
<tr><td colspan="2">Triggers</td><td>Working outside process. Unforeseen circumstances.</td></tr>
<tr><td colspan="2">Background</td><td>Storage is an area of activity in which everyone participates.
In some circumstances there are experts involved, designing and constructing customised solutions, but in many situations a storage scenario will be set by others. The challenge under CDM is for the design team to design storage facilities, arrangements and equipment to minimise risks to users.</td></tr>
<tr><td>E</td><td>Elimination</td><td>Significantly hazardous storage scenarios should be avoided SFARP.</td></tr>
<tr><td>R</td><td>Reduction</td><td>Risks should be minimised by informed choices.</td></tr>
<tr><td rowspan="2">I</td><td>Information for users</td><td>Informative notices and warning signs where appropriate.</td></tr>
<tr><td>Information for others</td><td>Instructions for maintenance and operation.</td></tr>
<tr><td>C</td><td>Controls envisaged</td><td>Signage.
Access equipment.
PPE as a last resort only (as it may be mislaid or not used for other reasons).</td></tr>
<tr><td colspan="2">Key references</td><td>• Cassie and Seale (2003)
• HSE (1998b, 1998c, 1998d, 2006d, 2007, 2009b, 2013i)
Statutes
• Dangerous Substances and Explosive Atmospheres Regulations 2002</td></tr>
<tr><td colspan="3">Notes</td></tr>
</table>

<table>
<tr><th colspan="3">Hazardous systems – radiation</th></tr>
<tr><td colspan="2">Risk</td><td>Cancer, blindness, radiation sickness, burns etc.</td></tr>
<tr><td colspan="2">Key features</td><td>Ionising and non-ionising radiation from planned or accidental exposure to radiation arising naturally or from human activity</td></tr>
<tr><td colspan="2">Triggers</td><td>Exposure to radiation in one of its many forms (see Background).
Potential sources of radiation:
• operation of equipment
• maintenance of equipment
• decommissioning and/or replacement of equipment.</td></tr>
<tr><td colspan="2">Background</td><td>The main difference between ionising and non-ionising radiation is the amount of energy the radiation carries, ie ionising radiation carries more energy.
Ionising radiation includes X-rays, gamma rays, radiation from radioactive sources and sources of naturally occurring radiation such as radon gas.
Ionising radiation has many uses in industry, such as energy production, manufacturing, medicine and research. Its benefits have to be accompanied by protection of people who may be exposed.
Non-ionising radiation includes visible light, ultraviolet light, infrared radiation and electromagnetic fields.
Ultraviolet light is part of natural sunlight and also forms part of some man-made light sources. It can cause a number of health problems including skin cancer.
Sources of electromagnetic fields are used extensively in telecommunications and manufacturing with little evidence of related long-term health problems.
Radon.</td></tr>
<tr><td>E</td><td>Elimination</td><td>Avoidance of exposure, particularly to the more harmful types of radiation</td></tr>
<tr><td>R</td><td>Reduction</td><td>Reduction of periods of exposure and levels of exposure
Planning to minimise the need for human exposure, such as the positioning of machinery
Isolation and/or screening of equipment, which is a potential source of radiation
Potential sources of radiation:
• provision of safe access routes and access around equipment for operation and maintenance
• design of safe systems of isolation.</td></tr>
<tr><td rowspan="2">I</td><td>Information for users</td><td>Clearly identify the risks using words and symbols, identify controlled areas and equipment, and provide safety instructions and access to management advice and expertise.</td></tr>
<tr><td>Information for others</td><td>Provide managers with in-depth technical information and access to specialist advice and expertise.</td></tr>
<tr><td>C</td><td>Controls envisaged</td><td>Compliance with current regulations and guidance.
Following of current best practice, consulting with relevant experts.
Protection from radon released from soils using barriers and vented spaces.
Protection, which may include physical separation, barriers, working methods etc (both in normal situations and during maintenance and other interventions).</td></tr>
<tr><td colspan="2">Key references</td><td>• HM Government (2004)
• HSE (2000b)
Statutes
• High-Activity Sealed Radioactive Sources and Orphan Sources (HASS) Regulations 2005
• Radioactive Substances Act 1993
• Radiation (Emergency Preparedness and Public Information) Regulations 2001
• Ionising Radiations Regulations 1999
• Ionising Radiations (Medical Exposure) Regulations 2000
Websites
• BRE Radon: www.bre.co.uk/page.jsp?id=3133
• Gov.uk: https://www.gov.uk/health-protection/radiation
• International commission on non-ionising radiation protection (ICNIRP)
• UKradon: http://ukradon.org/information</td></tr>
<tr><td colspan="3">Notes</td></tr>
</table>

<table>
<tr><th colspan="3">Normal activities – posture and manual handling</th></tr>
<tr><td colspan="2">Risk</td><td>Musculoskeletal injuries and complaints.</td></tr>
<tr><td colspan="2">Key features</td><td>Arising from poor posture, awkward movements, excessive loads that put a strain on the human body.
Normally accumulative and can lead to long-term ill-health.
Excessive repetition of simple movements may trigger repetitive strain injury (RSI), eg using computers (display screen equipment).
Difficult physical tasks may be associated with slips, trips and falls due to loss of balance and/or concentration.
Excessive carrying distances increases risk of grip being lost so that loads shift or are dropped, spilled or splashed.
Usually associated with discomfort, but not necessarily (eg RSI, which can become severe with little warning).</td></tr>
<tr><td colspan="2">Triggers</td><td>Repeated activity involving poor posture, awkward movements, excessive loads, bad 'fit' between people and the workplace and/or excessive repetition.</td></tr>
<tr><td colspan="2">Background</td><td>Rarely seriously considered until recent times. May be associated with other hazards. Ergonomics is a specialist subject, but most injuries and ill-health arise from common activities that are well-explained in HSE guidance, particularly carrying excessive weights incorrectly. Design of work areas should reflect functionality and fit, for example, kitchens should be designed with easily-accessed storage and lay-down areas for hot/heavy pans etc immediately next to hobs and ovens, and there should be adequate circulation space.</td></tr>
<tr><td>E</td><td>Elimination</td><td>Tasks may be changed and some may be mechanised.
Designers should consider access, loadings etc required to facilitate the use of mechanised lifting and handling equipment.
If the workplace and tasks are well-designed, the risk may be eliminated.</td></tr>
<tr><td>R</td><td>Reduction</td><td>Levels of risk may be reduced by reducing loads and frequency or by providing equipment.</td></tr>
<tr><td rowspan="2">I</td><td>Information for users</td><td>Informative notices.</td></tr>
<tr><td>Information for others</td><td>An outline policy for ergonomics may explain the approach assumed by designers. Note that workplace management will carry out their own in-use assessments and the policy may be amended.
For particular inherent workplace activities, information on limitations should be provided to management.</td></tr>
<tr><td>C</td><td>Controls envisaged</td><td>Mechanical aids, platforms etc to make tasks easier.
Manual handling risk assessments. Health surveillance. Training.</td></tr>
<tr><td colspan="2">Key references</td><td>• HSE (1983, 2002c, 2000d 2003a, 2003c, 2004, 2013i, 2013j, 2013k, 2014b)
Statutes
• Health and Safety (Display Screen Equipment) Regulations 1992
• Lifting Operations and Lifting Equipment Regulations (LOLER) 1998
• Manual handling Operations Regulations 1992</td></tr>
<tr><td colspan="3">Notes</td></tr>
</table>

Normal activities – use of vehicles		
Risk		Collision and trapping of drivers and people by vehicles, including forklift trucks and robotic transfer vehicles.
Key features		Death or injury.
Triggers		Operator error, impaired sight-lines, toppling over, other accidents etc.
Background		The Building Regulations Part K deal with some matters relating to vehicles and the Workplace Regulations 1992 also contain wide-ranging detailed requirements. Designers also need to consider risks that arise when vehicles will be in use beyond workplace premises (eg emptying gullies or during other maintenance activity).
E	*Elimination*	Risks should be removed SFARP, preferably by removing people from dangerous situations, ie segregation.
R	*Reduction*	Less risky options may exist and should be considered. Risks may be reduced, eg by providing clearly-defined pedestrian routes or by adding fail-safe controls.
I	*Information for users*	Advisory notices.
	Information for others	Circulation strategy in the O&M manual.
C	*Controls envisaged*	Various controls such as guards, warning signs, signals, horns etc may be appropriate.
Key references		• Evans (2005) • HM Government (2013b) • HSE (1992 [withdrawn], 1996a, 2000e, 2013a, 2013l, 2013s) *Statutes* • Workplace (Health, Safety and Welfare) Regulations 1992
Notes		

<table>
<tr><th colspan="3">Normal activities – use of plant and equipment</th></tr>
<tr><td colspan="2">Risk</td><td>A wide range of hazardous activity involving the use of blades, presses and other machinery, vehicles (see also above), access equipment, lifting equipment etc. Risks such as radiation arising from the use of specific equipment.</td></tr>
<tr><td colspan="2">Key features</td><td>Each activity carries risks both in respect of normal activity and situations that can arise when people have accidents or forget to wear PPE etc.</td></tr>
<tr><td colspan="2">Triggers</td><td>Equipment malfunction, abuse, accident etc.</td></tr>
<tr><td colspan="2">Background</td><td>The use of plant and equipment is legislated for in some areas of activity, but in every situation the risks need to be identified and assessed.
The Lifting Operations and Lifting Equipment (LOLER) Regulations 1998 relate to lifting activities and the Provision and Use of Work Equipment (PUWER) Regulations 1998 relate to equipment in general.</td></tr>
<tr><td>E</td><td>Elimination</td><td>Risks should be removed SFARP, preferably by removing/isolating people from dangerous situations.</td></tr>
<tr><td>R</td><td>Reduction</td><td>Less risky options may exist and should be considered. Risks may be reduced, eg by adding fail-safe controls.</td></tr>
<tr><td rowspan="2">I</td><td>Information for users</td><td>Advisory notices.</td></tr>
<tr><td>Information for others</td><td>Normally provided within the O&M manual.</td></tr>
<tr><td>C</td><td>Controls envisaged</td><td>Various controls such as guards, PPE, warning signs etc may be appropriate.</td></tr>
<tr><td colspan="2">Key references</td><td>• HSE (1996e, 2013i)
Statutes
• Lifting Operations and Lifting Equipment Regulations (LOLER) 1998
• Provision and Use of Work Equipment Regulations (PUWER) 1998
• Dangerous Substances and Explosive Atmospheres Regulations 2002
Note: the HSE has published a wide range of guidance notes on this subject on its website.
See also the section on radiation</td></tr>
<tr><td colspan="3">Notes</td></tr>
</table>

Normal activities – industrial processes		
Risk		There are risks inherent in some industrial processes.
Key features		Each activity carries risks both in respect of normal activity and situations which can arise when the process goes out of control.
Triggers		Events such as fatigue, corrosion, unstable reactions, contaminated materials, fluid leakage etc. Also explosions (see *Abnormal events - exposions*).
Background		These risks may be managed as part of the operation, but design of a new or adapted structure may provide opportunities to reduce overall risks.
E	*Elimination*	Hazardous processes should be avoided SFARP.
R	*Reduction*	Less risky options may exist and should be considered. Risks may be reduced, eg by adding fail-safe controls.
I	*Information for users*	Alarms, advisory notices, safe escape routes, PPE.
	Information for others	Normally provided within the O&M manual. May include outputs from HAZAN/HAZOP studies.
C	*Controls envisaged*	Various controls such as monitoring systems may be appropriate.
Key references		• HSE (1996b, 2014c, 2014d) ***Statutes*** • Control of Major Accident Hazards Regulations (COMAH) 1999 Note: the HSE has published a wide range of guidance notes on this subject on its website.
Notes		

Normal activities – working on highways		
Risk		In addition to the general risks of construction work the risks when working on highways include (a) workers being hit by vehicles and (b) creating hazards for road users.
Key features		Work near to road users.
Triggers		The need for routine and planned inspections, maintenance, major structure or equipment removal and/or replacement. Emergency workers and others dealing with accidents and incidents on the highway.
Background		Cost of management of safety during work on highways can be a significant part of the total cost. Much work is carried out at night. Safe management of machinery is a particular concern. Changes in system present particular traffic management risks.
E	*Elimination*	Specify materials that will avoid the need for their replacement or maintenance during the design life of the highway. Design to facilitate safe access for the inspection, maintenance and cleaning of structures and equipment including safe stopping places for maintenance vehicles, avoiding the need for maintenance workers to cross the carriageway. Design in remote monitoring equipment. Design to avoid creating risks for road users – carry out road safety audits assessing alignment, visibility, junction layout and visibility, non-motorised user provision, parking, loading/unloading facilities, road signs, carriageway markings and lighting. Avoid designing in features that could distract road users.
R	*Reduction*	Design to reduce the need for maintenance, the number of operatives needed and the duration of maintenance operations. Design to take account of the need for the installation of future traffic management equipment/facilities.
I	*Information for users*	Safe signing and speed control is vital.
	Information for others	Identify designed-in safety measures and assumptions in a designers' maintenance strategy document for inclusion in the H&S file.
C	*Controls envisaged*	Compliance with Highways England standards and guidance for carrying out works and DfT *et al* (2009a and b).
Key references		• HA (1997) • DfT, HA, DRDNI, TS, WAG (2009a and b)
Notes		

Normal activities – use of doors and windows/glazing		
Risk		Use of doors and windows, skylights and ventilators, and brittle materials.
Key features		May cause impact damage, trapping of fingers, cuts if glass breaks, falls from height.
Triggers		Accidents, inconsiderate use, horseplay. Presence of fragile materials in roof lights.
Background		Safe doors and windows are well covered in the Building Regulations Part K and the Workplace Regulations 1992. Glass (and its manifestation) is covered by the Building Regulations Parts M and N. Particular care is needed in buildings used by children and young people. Issues arising during refurbishment require particular attention to provide modern levels of protection.
E	***Elimination***	Avoid the specification of fragile material SFARP. Ensure compliance with Building Regulation requirements.
R	***Reduction***	Substitution with less fragile materials may be appropriate. Reduction of risks SFARP is always necessary. Use of toughened glass or (better still) laminated glass
I	***Information for users***	Users may need to be alerted to risks of impact as doors open or where glass is used.
	Information for others	Normally provided within the O&M manual.
C	***Controls envisaged***	Window opening may need to be restricted at height and hinges may need to be covered in schools or other places where there may be horseplay.
Key references		• Keiller *et al* (2005) • HM Government (1998, 2004b, 2013b) • HSE (2013a) ***Statutes*** • BS 8213-1:2004 • Workplace (Health, Safety and Welfare) Regulations 1992

Notes

Normal activities – use of lifts, escalators and moving walkways		
Risk		Use of lifts, escalators and moving walkways.
Key features		May cause impact damage, trapping of fingers. Detailed requirements are given in British Standards.
Triggers		Accidents, inconsiderate use, horseplay, breakdown, vandalism.
Background		Lifts are covered in the Building Regulations Part M and provision of escalators and moving walkways is included in the Workplace Regulations 1992. Particular care is needed in buildings used by children and young people.
E	*Elimination*	Not usually appropriate.
R	*Reduction*	Compliance with appropriate British Standards and industry best practice should ensure that installations are very safe to use, subject to regular inspection and maintenance.
I	*Information for users*	Advisory notices for users.
	Information for others	Normally provided within the O&M manual, especially with respect to inspection and maintenance requirements.
C	*Controls envisaged*	Cut-outs should be provided to enable users to cut off power in the event of an accident. Lifts should be easily accessed and well-signed to encourage their use rather than the use of escalators by those with heavy baggage.
Key references		• CIBSE (2010a) • HM Government (2004b) • Gilbertson (2013a and 2013b) • Page and Hough (1989) ***Statutes*** • BS 5656-2:2004
Notes		

Slips and trips while in motion on floors and ramps		
Risk		Accidental slipping or tripping due to a variety of contributory factors.
Key features		Injury and possibly death (either immediately or later).
Triggers		Smooth or polished surfaces, presence of contaminants including water, trip risks, inappropriate footwear, pedestrian behaviour, visual and noise distraction, trailing leads etc inappropriate or insufficient cleaning, low lighting levels or glare, or smooth flooring adjacent to safe flooring.
Background		Clear requirements are given in the Workplace Regulations 1992 and the HSE provides advice, including sponsoring the guidance by Lazarus *et al* (2006). The Building Regulations Parts K and M (HM Government, 2004b and 2013b) refer to ramps.
E	***Elimination***	By good design, the risk can be eliminated or reduced considerably, eg provision of safety flooring in predictably wet areas, adequate electrical and IT sockets to avoid trailing leads. Close attention to building tolerances especially at expansion joints, control joints and thresholds. Specify high-quality floor boxes that will not distort and present a trip risk. Ramps are higher risk than flat floors and need to be treated carefully with good visual contrast.
R	***Reduction***	Lower risk floor finishes should always be used. For a given floor type, risks may be reduced by good management and attention to the trigger features listed above.
I	***Information for users***	Visual signing by colour change, especially for ramps.
	Information for others	Cleaning and maintenance requirements should be in the O&M manual.
C	***Controls envisaged***	Keep flooring in good low-slip-risk condition, by keeping dry and free from trip risks. Manage cables and hoses.
Key references		• HM Government (2004b, 2013b) • HSE (2012d, 2013a) • Lazarus *et al* (2006) ***Statutes*** • Workplace (Health, Safety and Welfare) Regulations 1992 ***Websites*** • Centre for Accessible Environments (CAE): **http://cae.org.uk** • HSE *Slips and trips*: **www.hse.gov.uk/slips** • HSE *Slips assessment tool*: **www.hse.gov.uk/slips/sat**
Notes		

Slips and trips while using stairs and escalators		
Risk		Falls while negotiating stairs.
Key features		Injury and possibly death (either immediately or later).
Triggers		As for slips and trips (slippery surface, presence of contaminants including water, trip risks, inappropriate footwear, pedestrian behaviour, visual and noise distraction, trailing leads etc), but note that the consequences of falling on stairs are greater.
Background		Clear requirements are given in the Workplace Regulations 1992 and the Building Regulations Parts K and M (HM Government, 2004b, 2013b). Advice generally as for floors and ramps, but noting the greater risks on stairs.
E	*Elimination*	Consider ramps instead, where possible.
R	*Reduction*	Escalators or lifts may be considered. Stairs should be designed to be as safe as possible.
I	*Information for users*	Clearly visible tread edges.
	Information for others	Cleaning and maintenance requirements should be in the O&M manual.
C	*Controls envisaged*	Provision of handrails (mandatory under the Building Regulations).
Key references		• Gilbertson (2013a and 2013b) • HM Government (2004b, 2013b) • HSE (2012d, 2013a) • Lazarus *et al* (2006) ***Statutes*** • BS 5395 • BS 4211:2005+A1:2008 • BS 4592 • Workplace (Health, Safety and Welfare) Regulations 1992 ***Websites*** • HSE *Slips and trips*: **www.hse.gov.uk/slips** • HSE *Slips assessment tool*: **www.hse.gov.uk/slips/sat**
Notes		

<table>
<tr><th colspan="3">Slips and trips while essentially static, eg at a workstation, in a kitchen, using a bathroom</th></tr>
<tr><td colspan="2">Risk</td><td>Falls while moving around a room.</td></tr>
<tr><td colspan="2">Key features</td><td>Injury or possibly death – particularly if there are hot surfaces or hot liquids present.</td></tr>
<tr><td colspan="2">Triggers</td><td>As for slips and trips (slippery surface, presence of contaminants including water, trip risks, inappropriate footwear, pedestrian behaviour, visual and noise distraction, trailing leads etc), but note that the risks are often likely to be influenced by contamination from various spilt materials including liquids and powders, which reduce the slip resistance of a surface considerably.
In addition, the need for stretching and other physical movements such as manual handling may contribute to a slip.</td></tr>
<tr><td colspan="2">Background</td><td>Clear requirements are given in the Workplace Regulations 1992.</td></tr>
<tr><td>E</td><td>Elimination</td><td>Not possible.</td></tr>
<tr><td>R</td><td>Reduction</td><td>For a given floor type, risks may be reduced by good management and attention to all Trigger features.
Electrical sockets and storage should be easily reached without leaning over, eg hot surfaces or boiling pans.
Storage should be easily accessed without excessive stretching, particularly for heavy equipment.</td></tr>
<tr><td rowspan="2">I</td><td>Information for users</td><td>Advisory notices.</td></tr>
<tr><td>Information for others</td><td>Appropriate cleaning and maintenance requirements should be set down clearly for all spaces.</td></tr>
<tr><td>C</td><td>Controls envisaged</td><td>Keep flooring in good low-slip-risk condition. Tidy workplace.</td></tr>
<tr><td colspan="2">Key references</td><td>• Gilbertson (2013a and 2013b)
• HM Government (2004b, 2013b)
• HSE (2012d, 2013a)
• Lazarus et al (2006)
Statutes
• Workplace (Health, Safety and Welfare) Regulations 1992
Websites
• HSE Slips and trips: www.hse.gov.uk/slips
• HSE Slips assessment tool: www.hse.gov.uk/slips/sat</td></tr>
<tr><td colspan="3">Notes</td></tr>
</table>

Working at height – using equipment such as ladders		
Risk		Access to work at height.
Key features		Risk of fall that may cause injury or death.
Triggers		Unstable equipment or foundation, unsafe working practices such as poor foundation, leaning sideways and over-reaching, carrying materials, using tools, being struck, or by pure accident/misjudgement.
Background		Use of ladders is inherently risky, but their use is entrenched in 'normal practice' in particular due to the number of tasks that have been traditionally accomplished using ladders. Other equipment such as towers may appear safer, but their erection can be risky. All access to work at height is likely to present risks.
E	*Elimination*	SFARP design out the need for regular cleaning, maintenance and inspection that would necessitate work at height. Safer ways of meeting a need without needing to work at height should be designed in SFARP, eg by providing access platforms.
R	*Reduction*	If access to work at height is inescapable, safer means of access than ladders will frequently be appropriate. If ladders are to be used, appropriate provision for their safe use should be designed in, ie a safe working system provided for with accessible securing points.
I	*Information for users*	Advisory notices.
	Information for others	Strategy and provision to be explained within the O&M manual.
C	*Controls envisaged*	Fixed ladders, special access equipment such as cherry-pickers. Latch-on safety systems may be appropriate but they have their own dangers.
Key references		• Gilbertson (2013a and 2013b) • HM Government (2004b, 2013b) • HSE (2012d, 2013a, 2014e) • Lazarus *et al* (2006) ***Statutes*** • Work at Height Regulations 2005 ***Websites*** • HSE *Falls from height*: **www.hse.gov.uk/food/falls.htm** • Working at Height Safety Association: **www.wahsa.org.uk**
Notes		

Working at height – at unprotected edges and adjacent to fragile surfaces		
Risk		Access to work at height.
Key features		Risk of fall that may cause injury or death.
Triggers		Loss of balance, tripping or slipping, being knocked or by pure accident/misjudgement.
Background		Building Regulations Part K (HM Government, 2013b) deal with the provision of guards and barriers. Working at unguarded edges of floors/roofs is inherently risky, but is currently entrenched as 'normal practice' in particular due to the number of flat roofs without edge protection. The use of fragile roofing materials (including in skylights) is now reducing for new build, but many remain in existing structures and their dangers are often not appreciated.
E	*Elimination*	Design in permanent edge protection SFARP. Safer ways of meeting a need without needing to work at height at unprotected edges should be designed in SFARP.
R	*Reduction*	If access to work at height at unprotected edges is inescapable, provision may be made for fixing temporary edge protection, remembering that the fixing of such a system should not be a hazardous task. If work at height at unprotected edges is unavoidable, appropriate provision for safe working should be designed in, ie a safe working system provided for and explained.
I	*Information for users*	Advisory notices.
	Information for others	Strategy and provision to be explained within the O&M manual.
C	*Controls envisaged*	Latch-on safety systems.
Key references		• BRE (2005) • HM Government (2013b) • HSE (2014e) ***Statutes*** • BS 6180:2011 • Work at Height Regulations 2005 ***Websites*** • HSE *Falls from height*: **www.hse.gov.uk/food/falls.htm** • Working at Height Safety Association: **www.wahsa.org.uk**
Notes		

Abnormal events – fire		
Risk		Loss of control of combustion or spontaneous combustion of a structure or its contents.
Key features		Unpredictable and potentially fatal for all in a structure, including emergency services staff.
Triggers		Electrical fault, spark, naked flame, overheating, smoking, accident, arson, process malfunction, flammable liquid or solid present.
Background		For most types of building, the Building Regulations Part B (HM Government, 2006a) provide a sensible approach that is commonly adopted and accepted by the UK authorities. Alternative approaches may be taken using risk analysis and in all situations the approach should take account of particular circumstances that may lead designers to take precautions different from what is required by the Building Regulations. Provision of sprinkler protection should be actively considered, but taking account of planned/expected use of spaces. Detection systems should be considered taking account of the planned/expected use of spaces – smoke detectors may not work in dusty or steamy environments. Use of water to extinguish fires should be considered taking account of risks from electricity. The Regulatory Reform (Fire Safety) Order 2005 requires owners to do their own risk assessments (fire certificates now being obsolete) and designers would be well advised to consult with the party who will carry out the audit during the design, to reduce the risk of conflict when the audit is carried out.
E	*Elimination*	Remove triggers if possible. Remove flammable material: structure, stores, furniture and fittings, furnishings, decorative coatings, decorations.
R	*Reduction*	Use lower-risk processes. Use less flammable material: structure, stores, furniture and fittings, furnishings, decorative coatings, decorations. Design in fire-suppression systems. Design in alarm systems. Amend layout to permit faster escape and assist fire-fighting. Safer systems.
I	*Information for users*	Signage. Emergency communications (alarms, Tannoy systems). Fire-fighting instructions. Training of fire marshals.
	Information for others	Strategy statement. Plans of signage and escape routes. Restrictions on security measures. Maintenance regime.
C	*Controls envisaged*	Sprinkler systems. Fast identification and response. Smoke evacuation (preferably natural by chimney effect). Fast exit by clear, signed, smoke-free routes. Fire-fighting equipment.
Key references		• BRE (1996) • CIBSE (2010b) • DCLG (2006) • HM Government (2006a) • HSE (1995b, 1996b, 2013i, 2013n) ***Statutes*** • BS 9999:2008 • BS 5839-1:2013 • The Dangerous Substances and Explosive Atmospheres Regulations (DSEAR) 2002 • Regulatory Reform (Fire Safety) Order 2005 Note: the HSE has published a wide range of guidance notes on this subject on its website.
Notes		

Abnormal events – explosion		
Risk		Pressure wave and high-velocity missiles, flames and gases.
Key features		Causes immediate injury and death.
Triggers		Failure of a pressure vessel during testing or in use, loss of control of a process or stored material, ignition of gases or dusts, terrorism and war.
Background		Normally considered in particular industrial processes or experimental work or where hazardous materials (including gases) are employed or stored or when dust-clouds may occur.
E	*Elimination*	Risks should be removed SFARP, preferably by separating and protecting people from dangerous situations.
R	*Reduction*	Less risky options may exist and should be considered. Risks may be reduced, eg by adding fail-safe controls and alarms or providing vented enclosure. Special measures to avoid sparks where dust-clouds may occur.
I	*Information for users*	Advisory notices.
	Information for others	Normally provided within the O&M manual.
C	*Controls envisaged*	Various controls such as barriers, PPE, warning signs etc may be appropriate.
Key references		• HSE (2003b, 2013i, 2013n) ***Statutes*** • Dangerous Substances and Explosive Atmospheres Regulations 2002 • Directive 94/9/EC (the ATEX Directive) ***Websites*** • HSE *Dispensing petrol as a fuel*: **www.hse.gov.uk/pubns/indg216.htm** Note: the HSE has published a wide range of guidance notes on this subject on its website.
Notes		

<table>
<tr><th colspan="3">Abnormal events – falling objects</th></tr>
<tr><td colspan="2">Risk</td><td>Injury from falling objects.</td></tr>
<tr><td colspan="2">Key features</td><td>Cause injury or death.</td></tr>
<tr><td colspan="2">Triggers</td><td>Loss of control of material stored or being carried, or failure of overhead structures.</td></tr>
<tr><td colspan="2">Background</td><td>Proper control of stored materials is required, with planning and management to avoid misplacement, instability and overloading.
The Workplace Regulations 1992 provide sound guidance. Trees may present a risk.</td></tr>
<tr><td>E</td><td>Elimination</td><td>Protection should be provided where there is a risk. Ideally heavier materials should be stored on the ground.</td></tr>
<tr><td>R</td><td>Reduction</td><td>Lower risk options should be explored.
Risks may be reduced by design of storage systems and transport arrangements.</td></tr>
<tr><td rowspan="2">I</td><td>Information for users</td><td>Advisory notices and electronic reporting</td></tr>
<tr><td>Information for others</td><td>Storage and transportation strategy and details should be provided within the O&M manual</td></tr>
<tr><td>C</td><td>Controls envisaged</td><td>Advisory notices and electronic monitoring.</td></tr>
<tr><td colspan="2">Key references</td><td>• HM Government (2013b)
• HSE (2014e)
• Lonsdale (2013)
Statutes
• The Work at Height Regulations 2005</td></tr>
<tr><td colspan="3">Notes</td></tr>
</table>

Abnormal events – disproportionate collapse		
Risk		Structural collapse of a severe nature due to failure of critical members due to a local event.
Key features		Loss of life or injury due to partial or complete collapse of a structure.
Triggers		Impact (eg from a vehicle), explosion (eg from gas), failure of a key member (eg due to fatigue or corrosion).
Background		Building Regulations Part A (HM Government, 2004c) requires that this risk is considered in occupied buildings. Class A and B buildings may follow a prescribed route that may be appropriate for some situations, otherwise a risk assessment approach is required.
E	*Elimination*	Some risks can be removed by, for example, protecting a building from impact or not using gas.
R	*Reduction*	Selecting a framing system that has load-spreading ability should a key member fail will reduce the risks. Key members must be well tied into the building, and they may also be over-sized.
I	*Information for users*	Not applicable.
	Information for others	The approach taken, the evaluation of risks and the design precautions taken should be recorded with the structural design information passed to the owner.
C	*Controls envisaged*	Requirements for inspection and maintenance should be clearly recorded.
Key references		• BDA/AACPA/CPA (2015) • Cormie (2013) • HM Government (2004c) • IStructE (2010) • NHBC (2011) • Way (2004)
Notes		

Abnormal events – drowning and asphyxiation		
Risk		Drowning and asphyxiation of people.
Key features		Loss of life due to failure to breathe, either instantaneously or over time as the lungs are crushed.
Triggers		Falling into water or powders etc or being buried by collapsing materials. Flooding. Entry into confined spaces that may not support life. Release of gases from gas suppression systems for fire control.
Background		There are a range of risks such as falling into vats of water or powders, or being buried by grain etc during cleaning or trying to release blockages. Often, the challenge is to stop operatives entering dangerous situations. Designers need to consider drowning risks (especially risks to children) where any water is involved. Risk of drowning during a flood event should be considered where appropriate, in particular where sudden inundation could trap people who might not be able to escape in time from, for example, a basement, manholes, sewers etc.
E	***Elimination***	Where possible, people should be separated from such risks. Design in suitable ventilation of potentially confined spaces SFARP. Design to avoid the need for anyone to access any confined spaces SFARP. Design in suitable edge protection to areas of water, fluids etc SFARP.
R	***Reduction***	Controls to be provided where possible. Consider adequacy of access arrangements. Design in rescue equipment/system.
I	***Information for users***	Advisory notices, eg *'Do not enter silo unless empty – you may be buried and die – the grain will suck you under'.*
	Information for others	Separation and escape strategy should be provided within the O&M manual.
C	***Controls envisaged***	Confined spaces designed as securable/lockable environments with access controlled under permit/authority system. Provision of PPE and developed emergency escape system if entry is vital.
Key references		• HSE (2013m, 2014a) ***Statutes*** • The Confined Spaces Regulations 1997
Notes		

Abnormal events – overcrowding		
Risk		Large numbers of people behaving as a crowd.
Key features		Normally the behaviour is self-controlled, but in some circumstances such as panic, crush or escape, crowd behaviour can present a risk.
Triggers		Large numbers of people gathered in one space, ie an event that causes them to lose normal inhibitions and act as a group, normally for self-preservation, but sometimes for other purposes such as excitement.
Background		Crowd behaviour is a phenomenon that is experienced frequently in some places such as sports events, but can occur elsewhere such as in nightclubs. Advice on strength of barriers is provided in the Building Regulations Part K (HM Government, 2013b). Crowd flow may be modelled in a computer.
E	***Elimination***	Crowd behaviour cannot be entirely avoided.
R	***Reduction***	Crowds are less prone to uncontrolled behaviour if they are restrained in smaller groups and if the triggers are managed out.
I	***Information for users***	Signage needs to be easily seen when in a crowd. Signs may not be read, so a good public address system may enable a crowd to be managed positively, overcoming rumour spread by word of mouth.
	Information for others	A crowd management strategy should be provided in the O&M manual.
C	***Controls envisaged***	Good signage. Barriers are used to break up crowd packing and movement. Planning crowd escape routes along level surfaces without steps or street furniture etc will reduce accidents. Rapid easy egress from spaces should be provided, eg emergency escape doors.
Key references		• DCMS (2008) • HM Government (2013b) • HSE (2014f) • Rowe and Ancliffe (2009) • RSSB (2009)
Notes		

Abnormal events – responding to emergency events		
Risk		People responding to emergency events may be faced with risks that could be made less severe by designer risk management.
Key features		Events such as flooding, landslip, trees falling, damage to property, vehicle crashes, structural failure, fire, explosion.
Triggers		Exposure to dangers while dealing with the emergency events, arising due to weather conditions, accidents, structural failure etc.
Background		The issues to be considered will be wide-ranging.
E	*Elimination*	Not possible.
R	*Reduction*	Design structures to minimise the potential risks.
I	*Information for users*	Access to information and advice. Preparation through the existence of emergency plans. Training.
	Information for others	Access to information. Ability to provide information and advice to people on the ground.
C	*Controls envisaged*	Incorporation in a design of safety equipment such as guarding to prevent a worker being sucked into a drain that he/she is trying to unblock. Provision of accessible equipment and PPE.
Key references		• Jackson *et al* (2002)
Notes		

<table>
<tr><th colspan="3">Abnormal events – malicious human intervention</th></tr>
<tr><td colspan="2">Risk</td><td>Terrorism, vandalism, crime, arson, illicit use of drugs including tobacco etc.
Abusive or violent people.
Breaches of security, break-ins.</td></tr>
<tr><td colspan="2">Key features</td><td>Unpredictable and widely varying in scope and scale, leading to a variety of risks.</td></tr>
<tr><td colspan="2">Triggers</td><td>Malicious acts.</td></tr>
<tr><td colspan="2">Background</td><td>The issues to be considered are many and varied.</td></tr>
<tr><td>E</td><td>Elimination</td><td>Not entirely possible but certain security measures may be taken to reduce the chance of an occurrence.</td></tr>
<tr><td>R</td><td>Reduction</td><td>Not certain but certain defensive measures may be taken to reduce the impact of an occurrence.</td></tr>
<tr><td rowspan="2">I</td><td>Information for users</td><td>Not normally applicable.</td></tr>
<tr><td>Information for others</td><td>Possibly in a confidential document held by particular senior staff.</td></tr>
<tr><td>C</td><td>Controls envisaged</td><td>Physical separation.
Security coverage, automated alarms, planned response.
Additional levels of structural redundancy.
Avoidance of ‘attractors’ such as obvious opportunities for access, high-value accessible fittings etc.</td></tr>
<tr><td colspan="2">Key references</td><td>• Cheetham (1994)
• Clarke and Gilbertson (2011)
Websites
• Secured by design: www.securedbydesign.com</td></tr>
<tr><td colspan="3">Notes</td></tr>
</table>

4 Tables of information about typical workplaces

This chapter explores some of the workplace issues that arise in a small selection of the many thousands of different workplaces that exist.

The examples given are not meant to be exhaustive or general, but are indicative only. Designers of each individual structure will have to consider the various potential risks that exist in their project and deal with them accordingly.

The following workplaces are discussed in this chapter:

1 City-centre office block.
2 Maternity hospital.
3 Prison cell block.
4 Stretch of motorway.
5 Infant school.
6 Food processing plant.

Table 4.1 City centre office block

Features	Their implications	Discussion
Normal working conditions within the block are low-risk.		While an office is, at first sight, a low-risk environment, it should be borne in mind that there will be a wide range of behaviours exhibited. The population within will be fluid and any training will soon be forgotten. The behaviour of visitors may be uncontrolled. Maintenance activity will be an ongoing issue as it may affect people generally and certainly the resident contacts for access arrangements. Risks will be minimised if maintenance tasks are easy to access and carry out. Access for replacement of plant and equipment should be considered. The need for access to roofs should be minimised. Tasks such as gutter clearance made simple and safe by careful planning of the roofscape so that the need for access beyond safety railings is rarely necessary. Security issues should be carefully addressed so that the risk of unauthorised access to roofs and other hazardous areas is controlled.
Specific risks may exist that arise from the aspects of cleaning and maintenance that, if not designed for, may affect those who work in and visit the building.	Need to consider how all cleaning and maintenance can be achieved in a safe manner.	
Fire risks (and other related risks such as terrorism) will need careful consideration and evacuation procedures designed for.	How can large numbers of people be safely evacuated? Where can they safely be accommodated without risk from traffic?	
Water stored or used externally, eg in a cooling tower must be designed to minimise risk of Legionnaire's disease from air-borne sprays.	Reliance upon maintenance activity should be avoided if possible as there is always a risk of error or omission.	
Inevitably there will be areas that are hazardous to people if they can gain access.	Access to roofs and other hazardous areas will be abused if it is possible.	
Areas accessed only for maintenance may be seen as of secondary importance.	If access to plant-rooms etc is difficult, there will be a tendency to skimp on the inspection/maintenance work.	

Sources of advice

- British Council for Offices (BCO) publications (including Strickland, 2005)

Notes

Table 4.2 ***Maternity hospital***

Features	Their implications	Discussion
All hospitals need to be designed taking account of the patient profile and the nature of the work undertaken. In a maternity hospital, mothers and staff will need safe surroundings.	Issues of cleanliness and contamination risks will be important. Risks such as slipping or tripping should also be carefully addressed.	Not only is a maternity hospital a high-risk medical environment, it should borne in mind that there will be a wide range of behaviours exhibited. The population within will be fluid and any training will soon be forgotten. The behaviour of patients and visitors will mainly be uncontrolled. So, fire escape routes need to be simple, direct and clearly marked, leading to a designated and suitable safe area. This is normally achieved through a safe zoning compartment system. Infection and security issues will have to be addressed from the start as an integral part of the design. If maintenance work is carried out on an existing facility that will remain in use during the work, particular attention should be paid to the potential risks of the release of hazardous dusts, fumes etc.
In the event of fire, babies and non-ambulatory mothers will need to be safely evacuated and there needs to be a detailed strategy for this.	The strategy needs to address issues such as security, which may compromise the response, and where mothers and their babies can be rapidly accommodated in a warm safe environment, as well as addressing any urgent medical issues.	
Stealing of babies will be a security issue.	Provision of security should not compromise fire escape.	
Increasingly, the fight against infection is central to the choice of materials and their cleaning regimes.	The use of inherently germ-resistant materials with good slip resistance should be considered, eg lino.	
In the UK, the Department of Health has detailed guidance that should be followed.	All relevant advice should be considered at the start of the design process.	

Sources of advice

- DoH has a series of technical memoranda on a wide range of subjects. Anyone designing a health service building should be familiar with them, for example DOH (2013b) and Hopkins (2014)
- CIBSE (2008)
- HM Government (2006b)

Websites

- HSE *Moving and handling in health and social care*: **www.hse.gov.uk/healthservices/moving-handling.htm**

Notes

Table 4.3 ***Prison cell block***

<table>
<tr><th>Features</th><th>Their implications</th><th>Discussion</th></tr>
<tr><td>Prisoners are locked in their cells.</td><td>Need to release prisoners in case of fire etc or make cells 'safe havens' capable of surviving events.</td><td rowspan="7">A prison cell is obviously a hazardous environment where as far as possible hazards need to be designed out and the level of risk reduced.
Conversely, the presence of trained staff, normally in control of the environment, enables many hazardous situations to be rapidly responded to and managed through.
The unpredictability and potential disruptive behaviour of the prisoners makes the situation more difficult.
The management of prisons is a highly skilled activity, which includes the provision to prisoners of a reliable, disciplined environment including earned rewards and respect. These contribute to management of the ever-present risks.
Careful consideration needs to be given to the location of plant rooms, services etc that will require periodic maintenance/inspection access.</td></tr>
<tr><td>Movement generally is hampered by security measures.</td><td>Can electrical systems be relied upon? Is there manual over-ride?</td></tr>
<tr><td>The environment is heavily managed at all times.</td><td>Unlike most places of work, incidents should be seen and responded to rapidly at all times.</td></tr>
<tr><td>Response measures can be planned and staff can be trained and practice response activity.</td><td>Unlike most places of work, response to incidents is a major part of the job for all staff.</td></tr>
<tr><td>Prisoners may cause or contribute to generally hazardous situations.</td><td rowspan="2">Unlike most places of work, there are large numbers of people present with time on their hands who are often inclined to cause or at least join in mayhem.</td></tr>
<tr><td>Prisoners may attack staff or other prisoners under cover of incidents.</td></tr>
<tr><td>Prisoners may self-harm or commit suicide.</td><td>Some prisoners may present a risk of irrational action, which is a hazard to themselves or others.</td></tr>
<tr><td colspan="3">Sources of advice
• HM Prison Service: https://www.gov.uk/government/organisations/hm-prison-service</td></tr>
<tr><td colspan="3">Notes</td></tr>
</table>

Table 4.4 *Stretch of motorway*

<table>
<tr><th>Features</th><th>Their implications</th><th>Discussion</th></tr>
<tr><td>Motorways are a workplace for maintenance staff, traffic police and those providing breakdown assistance.</td><td></td><td rowspan="4">Where decisions are driven by the client, then the client is a designer and responsible for the decision. In the case of highways, where the details of provision are set out in great detail, if a designer wishes to make a change after assessing the 'in-use' risks and the client refuses, it is the client's decision as a designer.

Many of the issues arising require a holistic approach with whole-life cost modelling to achieve sensible outcomes.</td></tr>
<tr><td>The UK Government constantly strives for safer roads and requirements for safe work during maintenance are laid down by the Highways England.</td><td></td></tr>
<tr><td>It is known that stopping on the hard shoulder is hazardous.</td><td>Designers should assess ideas such as:

whether the provision of periodic off-road laybys for repairs is reasonably practicable

whether the provision of rumble-strips at the edge of the hard shoulder would be effective.</td></tr>
<tr><td>It is known that working on live motorways – even with good controls – is hazardous.</td><td>Design to minimise the need for maintenance should be the aim, eg the use of stainless steel lamp-posts.

Designers should consider access requirements for plant, scaffolds etc for future cleaning, maintenance and, where necessary, replacement of permanent equipment, signs etc.</td></tr>
<tr><td colspan="3">Sources of advice
• HSE (2001b)
• Road Liaison Group (2004)
Websites
• Highways England Roadways/site traffic control: www.highways.gov.uk/traffic-information</td></tr>
<tr><td colspan="3">Notes</td></tr>
</table>

Table 4.5 Infant school

Features	Their implications	Discussion
The main feature is that young children need to be looked after.	Children should be separated form hazards SFARP, eg no ponds, controlled access onto roads, good security.	All aspects of the design of a school need to take account of the limitations of children, their possible behaviour and their security and safety. Both overall layout and the design of means of access from one space to another should be carefully considered to avoid hazards and minimise risk. The planning of external areas is also important as it will involve vehicle movements and security issues. Issues that are specific to children need to be considered, eg hinge protection, safe planting, rounded corners to walls and steps. Issues that apply to both adults and children need to be considered so that the needs of both are taken into account, eg heights of handles, floor finishes.
A school is a specialist environment, so while general designers may feel they understand it, they are unlikely to.	Designers of schools need to liaise with education experts concerning the real hazards experienced by children.	
Children may behave in an excitable manner.	Doors may be slammed trapping fingers in hinges. Children may rush out onto roads to meet their parents, so control of gates is vital. Children are liable to have frequent falls.	
Young children may be curious and yet have little understanding of danger.	Access by children to places where they may become trapped or be exposed to hazards (such as hot pipes, ovens, refrigerators etc) should be designed to be controllable. Likewise, plants that are poisonous to humans should not be present.	
Security of children is a major concern of all parents and teachers.	Prevention of unauthorised access to children should be provided.	
Teachers will wish to mount displays in the classroom and other spaces.	Suitable display systems may be designed, eg pulley-operated to avoid standing on desks.	
Parents will wish to deliver and collect their children by car.	Provision of adequate drop-off/ collection parking facilities. If on-site, provision of a special fenced area is required.	
Teachers will wish to park on-site.	Separation of teachers' vehicles from children is required.	

Sources of advice

- DfE has a series of building bulletins on a wide range of subjects. Anyone designing a school should be familiar with them, for example Hopkins (2014) and DfE (2007)
- DCLG (2006)

Notes

Table 4.6 Food processing plant

Features	Their implications	Discussion
Factories are a specialist workplace and employ a range of staff from shift workers, engineers, fitters, electricians, packers, drivers and office staff. Tasks range widely and involve all manner of hazards in the course of the day-to-day operation. However, many staff generally repeat similar actions or tasks. There can be a high turnover of staff and labour can be unskilled.	Design of production process is specialised and can be industrially sensitive. Factories constitute a high-risk environment particularly to those unfamiliar with the facility.	Design should be undertaken in close consultation with the client/end user to ensure the appropriate consideration is given to the management of risks. Productivity is a prime driver and training and/or controls can be abused to maximize throughput. Where possible the design should prohibit abuse of plant or control systems. Consideration is required to the ergonomics in design.
Cleanliness and control of contamination. Note that up to 80 per cent of all accidents in food processing plants are slips caused mainly by poor flooring choice, poor cleaning, gross contamination, poor footwear and condensation. Conditions are often aggressive to products, particularly services and finishes.	The use of suitable finishes and grades of material is essential. Lifespan of finishes need to be considered and maintenance considerations designed in. Once in operation access can be impossible until plant is de-commissioned.	All areas of the production facility should be accessible for cleaning or inspection, hollow sections, voids etc should be designed out. There should be no pathway for foreign matter into the process. Windows and doors should prevent either ingress or support to flora/fauna. Where not essential glass and timber should be designed out. Specified materials should be durable and the lifespan required identified. Clients input into performance requirements ideally required.
The layout of plant is driven by the process sequence and operation. The surrounding environment is dependent upon the nature of the process. This can require abnormally high or low temperature etc.	Production areas are frequently congested and confusing. The environment can be unpleasant and/or disorientating with particular reference to excessive heat, cold, noise, steam and or aroma.	Access for inspection, operation or maintenance including valves etc should be considered within the initial design. Fire escape routes should be clearly defined and maintained. The likely environment and its effect on materials (shrinkage, expansion, condensation etc) need to be known.
Hazardous chemicals. Moving machinery.	High risk to those unfamiliar with environment.	Access to particular plant areas/process should be controllable.

Sources of advice

- HM Government (2006c)

Notes

References

ACDP (2003) *Infection at work: controlling the risks,* Advisory Committee on Dangerous Pathogens, Health and Safety Executive, Norwich, UK. Go to: **www.hse.gov.uk/pubns/infection.pdf**

ACDP (2005) *Biological agents: managing the risk in laboratories and healthcare premises*, Advisory Committee on Dangerous Pathogens, Health and Safety Executive, Norwich, UK. Go to: **www.hse.gov.uk/biosafety/biologagents.pdf**

ARMSTRONG (2002) *Facilities management manuals. A best practice guide*, C581, CIRIA, London, UK (ISBN:978-0-86017-581-0). Go to: **www.ciria.org**

BDA, AACPA, CBA (2015) *Masonry design for disproportionate collapse requirements under Regulation A3 of the Building Regulations (England & Wales)*, Brick Development Association, Autoclaved Aerated Concrete Products Association, Concrete Block Association, UK-USA. Go to: **http://tinyurl.com/o5egqgc**

BP SAFETY GROUP (2004) *Hazards of steam*, BP Process Safety Series, Institution of Chemical Engineers, Rugby, UK (978-0-85295-468-3). Go to: **http://tinyurl.com/p3gwq8j**

BRE (1992) *Remedial wood preservatives: use them safely*, BRE Digest 371, BRE Press, Building Research Establishment, Garston, UK (ISBN: 0-85125-534-5)

BRE (1996) *Escape of disabled people from fire*, BR301, BRE Press, Building Research Establishment, Garston, UK (ISBN: 978-1-86081-067-1)

BRE (2005) *Safety considerations in designing roofs*, BRE Digest 493, BRE Press, Building Research Establishment, Garston, UK (ISBN: 978-1-86081-753-3)

BSRIA (1988) *Ventilation effectiveness in mechanical ventilation systems*, TN1/88, Building Services Research and Information Association, Bracknell, UK (ISBN: 978-0-86022-189-0)

CASSIE, S and SEALE, L (2003) *Chemical storage tank systems – good practice: guidance on design, manufacture, installation, operation, inspection and maintenance*, C598, CIRIA, London, UK (ISBN: 978-0-86017-598-8). Go to: **www.ciria.org**

CDC (2005) *NIOSH Pocket guide to hazardous chemicals*, National Institute for Occupational Safety and Health, Centers for Disease Control and Prevention, Atlanta, GA, USA. Go to: **www.cdc.gov/niosh/npg**

CHEETHAM, D W (1994) *Dealing with vandalism*, SP91, CIRIA, London, UK (ISBN: 978-0-86017-375-5). Go to: **www.ciria.org**

CIBSE (1983a) *Lighting in hostile and hazardous environments*, Chartered Institution of Building Services Engineers, London, UK (ISBN: 0-90095-326-8)

CIBSE (1986) *Guide K. Electricity in buildings*, Chartered Institution of Building Services Engineers, London, UK. Go to: **http://tinyurl.com/pjwvyw5**

CIBSE (1996) *Lighting guide: areas for visual display terminals, CIBSE Application guide*, Chartered Institution of Building Services Engineers, London, UK (ISBN: 978-0-90095-341-5)

CIBSE (2004) *Guide B: Heating, ventilating, air conditioning and refrigeration*, Chartered Institution of Building Services Engineers, London, UK. Go to: **http://tinyurl.com/ovm47oy**

CIBSE (2005) *Natural ventilation in non-domestic buildings*, AM10, Chartered Institution of Building Services Engineers, London, UK. Go to: **http://tinyurl.com/ncsxj5z**

CIBSE (2008) *Hospitals and health care buildings*, SLL LG2, Chartered Institution of Building Services Engineers, London, UK (ISBN: 978-190328-799-6). Go to: **http://tinyurl.com/q56oko8**

CIBSE (2010a) *Guide D: Transportation systems in buildings*, Chartered Institution of Building Services Engineers, London, UK. Go to: **http://tinyurl.com/ocsc8ag**

CIBSE (2010b) *Guide E: Fire safety engineering, third edition*, Chartered Institution of Building Services Engineers, London, UK. Go to: **http://tinyurl.com/nfvmadv**

CIBSE (2012a) *Code for interior lighting, revised*, SLL, Chartered Institution of Building Services Engineers, London, UK. Go to: **http://tinyurl.com/ncdvq9v**

CIBSE (2012b) *Lighting guide 01: the industrial environment*, SLL LGI, Chartered Institution of Building Services Engineers, London, UK. Go to: **http://tinyurl.com/okrflt9**

CIBSE (2013) *Minimising the risk of legionnaire's disease*, TM13, Chartered Institution of Building Services Engineers, London, UK. Go to: **http://tinyurl.com/q7fxjkq**

CIBSE (2014) *Guide G: Public health and plumbing engineering*, Chartered Institution of Building Services Engineers, London, UK. Go to: **http://tinyurl.com/p2tcq9a**

CIBSE (2015) *Guide A: Environmental design*, Chartered Institution of Building Services Engineers, London, UK (ISBN: 978-1-90328-766-8). Go to: **http://tinyurl.com/pzhosyx**

CIRIA (1999) *Operation and maintenance manuals for building – a guide to procurement and preparation*, C507, CIRIA, London (ISBN: 978-0-86017-507-0). Go to: **www.ciria.org**

CLARKE, L and GILBERTSON, A (2011) *Addressing crime and disorder in public places through planning and design*, C710, CIRIA, London, UK (ISBN: 978-0-86017-712-8). Go to: **www.ciria.org**

CORMIE, D (2013) *Manual for the systematic risk assessment of high risk structures against disproportionate collapse*, The Institution of Structural Engineers, London, IUK (ISBN: 978-1-90633-524-3)

DCLG (2006) *Fire safety risk assessment*, Department for Communities and Local Government Publications, West Yorkshire, UK (ISBN: 978-1-85112-819-8). Go to: **http://tinyurl.com/q8ue39n**

DCMS (2008) *Guide to safety at sports grounds*, Department for Culture, Media and Sport, HMSO, Norwich, UK (ISBN: 978-0-11702-074-0). Go to: **http://tinyurl.com/c583o9a**

DFE (2007) *Design for fire safety in schools*, Building Bulletin 100, RIBA Enterprises, London, UK. Go to: **http://tinyurl.com/pktx2g5**

DfT, HA, DRDNI, TS, WAG (2009a) "Traffic safety measures and signs for road works and temporary situations", *Traffic signs manual, Part 1 Design, Chapter 8*, The Stationery Office, HMSO, London, UK (ISBN: 978-0-11553-051-7). Go to: **http://tinyurl.com/poqgvd5**

DfT, HA, DRDNI, TS, WAG (2009b) "Traffic safety measures and signs for road works and temporary situations", *Traffic signs manual, Part 2 Operations, Chapter 8*, The Stationery Office, HMSO, London, UK (ISBN: 978-0-11553-052-4). Go to: **http://tinyurl.com/pfe49ay**

DOH (2010) *Specialist services. Health technical memorandum 08-02 Lifts*, Department of health, London, UK. Go to: **http://tinyurl.com/qxcwepk**

DOH (2013a) *Prevention and control of infection in care homes. An information resource*, Department for Health, London, UK. Go to: **http://tinyurl.com/p9l5r6r**

DOH (2013b) *Specialist services. Health technical memorandum 08-01 Acoustics*, Department of health, London, UK. Go to: **http://tinyurl.com/pczsbt6**

EI (1981) *Model code of safe practice in the petroleum industry. Part 3: refining safety code, third edition*, Energy Institute, London, UK (ISBN: 0-47126-196-3)

EI (1991) *Model code of safe practice in the petroleum industry. Part 1 Electrical safety code, revised*, Energy Institute, London, UK (ISBN: 0-47192-159-9)

EI (1998) *Model code of safe practice in the petroleum industry. Part 2: Design, construction and operation of distribution installations*, Energy Institute, London, UK (ISBN: 0-85293-204-9)

EVANS, G (2005) *HSE survey of first aid training organisations*, RR358, Health and Safety Executive, Sudbury, UK (ISBN: 0-71766-137-7). Go to: www.**hse.gov.uk/research/rrpdf/rr358.pdf**

GILBERTSON, A (ed) (2013a) Safer stairs in public places – assessment of existing stairs, C722, CIRIA, London (ISBN: 978-0-86017-725-8). Go to: **www.ciria.org**

GILBERTSON, A (ed) (2013b) *Safer escalators in public places*, C732, CIRIA, London, UK (ISBN: 978-0-86017-736-74). Go to: **www.ciria.org** (free download)

HA (1997) *Design manual for roads and bridges (DMRB)*, Highways England, UK. Go to: **www.standardsforhighways.co.uk/dmrb/**

HAWKINS, G (2011) *Rules of thumb. Guidelines for building services, fifth edition*, BSRIA Ltd, Berkshire, UK (ISBN: 978-0-86022-691-8)

HM GOVERNMENT (1992) *The Building Regulations 2000. Toxic substances. Approved Document D*, Office of the Deputy Prime Minister, London (ISBN: 978-1-85946-203-4). Go to: **http://tinyurl.com/nnkqzha**

HM GOVERNMENT (1998) *The Building Regulations 2000. Glazing – safety in relation to impact, opening and cleaning. Approved Document N*, Office of the Deputy Prime Minister, London (ISBN: 978-1-85946-212-6). Go to: **http://tinyurl.com/qfnaw3y**

HM GOVERNMENT (2003) *The Building Regulations 2010. Resistance to the passage of sound. Approved Document E*, Office of the Deputy Prime Minister, London (ISBN: 978-1-85946-204-1). Go to: **http://tinyurl.com/nhfej6b**

HM GOVERNMENT (2004a) *The Building Regulations 2010. Site preparation and resistance to contaminants and moisture. Approved Document C*, Office of the Deputy Prime Minister, London (ISBN: 978-1-85946-202-7). Go to: **http://tinyurl.com/qaoj6tn**

HM GOVERNMENT (2004b) The Building Regulations 2010. Access to and use of buildings. Approved Document M, Office of the Deputy Prime Minister, London (ISBN: 978-1-85946-211-9). Go to: **http://tinyurl.com/3c348g**

HM GOVERNMENT (2004c) *The Building Regulations 2000. Structure. Approved Document A*, Office of the Deputy Prime Minister, London (ISBN: 978-1-85946-200-3). Go to: **www.planningportal.gov.uk/uploads/br/BR_PDF_AD_A_2004.pdf**

HM GOVERNMENT (2006a) *The Building Regulations 2010. Fire safety. Volume 2 Buildings other than dwelling houses. Approved Document B*, Office of the Deputy Prime Minister, London (ISBN: 978-1-85946-262-1).
Go to: **www.planningportal.gov.uk/uploads/br/BR_App_Doc_B_v2.pdf**

HM GOVERNMENT (2006b) *Fire safety risk assessment – healthcare premises*, OPSI, Norwich, UK (ISBN: 978-1-85112-824-2). Go to: **http://tinyurl.com/oku66ew**

HM GOVERNMENT (2006c) *Fire safety risk assessment – factories and warehouses*, OPSI, Norwich, UK (ISBN: 978-1-85112-816-7). Go to: **http://tinyurl.com/oz63mm2**

HM GOVERNMENT (2010a) *The Building Regulations 2010. Ventilation. Approved Document F*, Office of the Deputy Prime Minister, London (ISBN: 978-1-85946-370-3). Go to: **www.planningportal.gov.uk/uploads/br/BR_PDF_AD_F_2010_V2.pdf**

HM GOVERNMENT (2010b) *The Building Regulations 2010. Sanitation, hot water safety and water efficiency. Approved Document G*, Office of the Deputy Prime Minister, London (ISBN: 978-1-85946-323-9). Go to: **http://tinyurl.com/m4c4z3f**

HM GOVERNMENT (2010c) *The Building Regulations 2000. Combustion appliances and fuel storage systems. Approved Document J*, Office of the Deputy Prime Minister, London (ISBN: 978-1-85946-371-0). Go to: **http://tinyurl.com/38qjk4u**

HM GOVERNMENT (2013a) *The Building Regulations 2000. Electrical safety – dwellings. Approved Document P*, Office of the Deputy Prime Minister, London (ISBN: 978-1-85946-485-4). Go to: **http://tinyurl.com/kanwftp**

HM GOVERNMENT (2013b) *The Building Regulations 2010.Protection from falling, collision and impact. Approved Document K*, Office of the Deputy Prime Minister, London (ISBN: 978-1-85946-484-7). Go to: **http://tinyurl.com/pncm3at**

HOPKINS, C, HALL, R, JAMES, A, ORLOWSKI, R, WISE, S and CANNING, D (2014) *Acoustic design of schools. A design guide*, Building Bulletin 93, Department for Education and Skills, London (ISBN: 0-11271-105-7).
Go to: **http://tinyurl.com/nnblmyc**

HSE (1980) *Container terminals: safe working practice*, Health and Safety Executive, Sudbury, UK (ISBN: 978-0-11883-302-8)

HSE (1983) *Suspended access euipment, revised, guidance note,* Health and Safety Executive, Sudbury, UK (ISBN: 978-0-11883-577-0)

HSE (1989) *Memorandum of guidance on the Electricity at Work Regulations 1989. Guidance on Regulations*, HSR25, Health and Safety Executive, Sudbury, UK (ISBN: 978-0-71766-228-9)

HSE (1992) *Road transport in factories and similar workplaces*, GS9(R), Health and Safety Executive, Sudbury, UK (ISBN: 978-0-11885-732-1)

HSE (1995a) *How to deal with sick building syndrome. Guidance for employers, building owners and building managers*, HSG132, Health and Safety Executive, Sudbury, UK (ISBN: 978-0-71760-861-4). Go to: **www.hse.gov.uk/pubns/priced/hsg132.pdf**

HSE (1995b) *Energetic and spontaneously combustible substances: Identification and safe handling*, HSG131, Health and Safety Executive, Sudbury, UK (ISBN: 978-0-71760-893-5). Go to: **http://tinyurl.com/oqqhd5z**

HSE (1995c) *Control of substances hazardous to health in the production of pottery*, L60, Health and Safety executive, Sudbury, UK (ISBN: 978-0-71760-849-2). Go to: **www.hse.gov.uk/pubns/priced/l60.pdf**

HSE (1996a) *Lift trucks in potentially flammable atmospheres*, HSG113, Health and Safety Executive, Sudbury, UK (ISBN: 978-0-71760-706-8)

HSE (1996b) *Safe use and handling of flammable liquids*, HSG140, Health and Safety Executive, Sudbury, UK (ISBN: 978-0-71760-967-3). Go to: **www.hse.gov.uk/pubns/priced/hsg140.pdf**

HSE (1996c) *A guide to the Work in Compressed Air Regulations 1996*, L96, Health and Safety Executive, Sudbury, UK (ISBN: 978-0-71761-120-1) (withdrawn)

HSE (1996e) *Health and safety in construction*, Health and Safety Executive, Sudbury, UK (ISBN: 978-0-71766-182-2).
Go to: **http://tinyurl.com/q5fdm6n**

HSE (1997a) *Lighting at work*, HSG38, Health and Safety Executive, Sudbury, UK (ISBN: 978-0-71761-232-1)

HSE (1997b) *Anthrax: safe working and the prevention of infection*, HSG174, Health and Safety Executive, Sudbury, UK (ISBN: 978-0-71761-415-8)

HSE (1997c) *Biological monitoring of the workplace*, HSG167, Health and Safety Executive, Sudbury, UK (ISBN: 978-0-71761-279-6)

HSE (1997d) *Further guidance on emergency plans for major accident hazard pipelines. The Pipeline Safety Regulations 1996*, H7/07, Health and Safety Executive, Sudbury, UK. Go to: **www.hse.gov.uk/pipelines/emergencyplanpipe.pdf**

HSE (1998a) *Noise in engineering*, EIS26, Health and Safety Executive, Sudbury, UK. Go to: **http://tinyurl.com/p2r9vg3**

HSE (1998b) *The storage of flammable liquids in containers*, HSG51, Health and Safety Executive, Sudbury, UK (ISBN: 978-0-71761-471-4)

HSE (1998c) *The storage of flammable liquids in tanks*, HSG176, Health and Safety Executive, Sudbury, UK (ISBN: 978-0-71761-470-7)

HSE (1998d) *The spraying of flammable liquids*, HSG178, Health and Safety Executive, Sudbury, UK (ISBN: 978-0-71761-483-7)

HSE (1999) *Health and safety in engineering workshops*, HSG129, Health and Safety Executive, Sudbury, UK (ISBN: 978-0-71761-717-3). Go to: **www.hse.gov.uk/pubns/priced/hsg129.pdf**

HSE (2000a) *General ventilation in the workplace: guidance for employers*, HSG202, Health and Safety Executive, Sudbury, UK (ISBN: 0-71761-793-9). Go to: **www.ucu.org.uk/media/pdf/f/g/HSG202_-_Ventilation.pdf**

HSE (2000b) *Work with ionising radiation. Ionising Radiations Regulations 1999. Approved code of practice and guidance*, L121, Health and Safety Executive, Sudbury, UK (ISBN: 978-0-71761-746-3). Go to: **www.hse.gov.uk/pubns/priced/l121.pdf**

HSE (2000c) *Protection of outside workers against ionising radiation*, IRP4, Health and Safety Executive, Sudbury, UK. Go to: **www.hse.gov.uk/pubns/irp4.pdf**

HSE (2000d) *Backs for the future – safe manual handling in construction*, HSG149, Health and Safety Executive, Sudbury, UK (ISBN: 0-71761-122-1)

HSE (2000e) *Safety in working with lift trucks*, HSG6, Health and Safety Executive, Sudbury, UK (ISBN: 978-0-71761-781-4)

HSE (2000f) *Management of health and safety at work. Management of Health and Safety at Work Regulations 1999. Approved code of practice and guidance, second edition*, L21, Health and Safety Executive, Sudbury, UK (ISBN: 978-0-71762-488-1). Go to: **http://tinyurl.com/qjrlxf5**

HSE (2001a) *The management, design and operation of microbiological containment laboratories*, Health and Safety Executive, Sudbury, UK (ISBN: 978-0-71762-034-0)

HSE (2001b) *Reducing at-work road traffic incidents, Work-related Road Safety Task Group*, Health and Safety Executive, Sudbury, UK (ISBN: 0-71762-239-8). Go to: **www.hse.gov.uk/roadsafety/experience/traffic1.pdf**

HSE (2002a) *Control of lead at work, third edition*, L132, Health and Safety Executive, Sudbury, UK (ISBN: 978-0-71762-565-9)

HSE (2002b) *Management of asbestos in buildings*, HSG227, Health and Safety Executive, Sudbury, UK (ISBN: 978-0-71762-381-5)

HSE (2002c) *Seating at work, third edition*, HSG57, Health and Safety Executive, Sudbury, UK (ISBN: 978-0-71761-231-4). Go to: **www.hse.gov.uk/pUbns/priced/hsg57.pdf**

HSE (2003a) *Work with display screen equipment: Health and Safety (Display Screen Equipment) Regulations 1992 as amended by the Health and Safety (Miscellaneous Amendments) Regulations 2002*, L26, Health and Safety Executive, Sudbury, UK (ISBN: 978-0-71762-582-6). Go to: **www.hse.gov.uk/pubns/priced/l26.pdf**

HSE (2003b) *Safe handling of combustible dusts: Precautions against explosions, second edition*, HSG103, Health and Safety Executive, Sudbury, UK (ISBN: 978-0-71762-726-4). Go to: **www.hse.gov.uk/pUbns/priced/hsg103.pdf**

HSE (2003c) *Safety in window cleaning using suspended and powdered access equipment*, MISC611, Health and Safety Executive, Sudbury, UK. Go to: **www.hse.gov.uk/pubns/misc611.pdf**

HSE (2004) *Manual handling. Manual Handling Operations Regulations 1992 (as amended)*, L23, Health and Safety Executive, Sudbury, UK (ISBN: 978-0-71762-823-0). Go to: **www.hse.gov.uk/pubns/priced/l23.pdf**

HSE (2005a) *Hand-arm vibration*, L140, Health and Safety Executive, Sudbury, UK (ISBN: 978-0-71766-125-1)

HSE (2005b) *Whole-body vibration*, L141, Health and Safety Executive, Sudbury, UK (ISBN: 978-0-71766-126-8). Go to: **www.hse.gov.uk/pubns/priced/l141.pdf**

HSE (2005c) *Controlling noise at work*, L108, Health and Safety Executive, Sudbury, UK (ISBN: 978-0-71766-164-0) Go to: **www.hse.gov.uk/pubns/priced/l108.pdf**

HSE (2005d) *Personal protective equipment at work (second edition)*, L25, Health and Safety Executive, Sudbury, UK (ISBN: 978-0-71766-139-8) Go to: **www.hse.gov.uk/pubns/priced/l25.pdf**

HSE (2006a) *COSHH essentials for welding, hot work and allied processes. Surface preparation: pressure blasting (large items)*, WL20, Health and Safety Executive, Sudbury, UK. Go to: **www.hse.gov.uk/pubns/guidance/wl20.pdf**

HSE (2006b) *The safe isolation of plant and equipment*, HSG253, Health and Safety Executive, Sudbury, UK (ISBN: 978-0-71766-171-8)

HSE (2006c) *Monitoring strategies for toxic substances, second edition*, HSG173, Health and Safety Executive, Sudbury, UK (ISBN: 978-0-71766-188-6)

HSE (2006d) *Handling and stacking of bales*, INDG125(rev2), Health and Safety Executive, Sudbury, UK. Go to: **www.hse.gov.uk/pubns/indg125.pdf**

HSE (2006e) *A guide to the Control of Major Accident Hazards Regulations 1999 (as amended)*, L111, Health and Safety Executive, Sudbury, UK (ISBN: 978-0-71766-175-6). Go to: **www.hse.gov.uk/pubns/priced/l111.pdf**

HSE (2006f) *Essentials of health and safety at work, fourth edition*, Health and Safety Executive, Sudbury, UK (ISBN: 978-0-71766-179-4)

HSE (2007) *Warehousing and storage: a guide to health and safety*, HSG76, Health and Safety Executive, Sudbury, UK (ISBN: 978-0-71766-225-8). Go to: **www.hse.gov.uk/pubns/priced/hsg76.pdf**

HSE (2009a) *COSHH essentials: easy steps to control chemicals*, HSG193, Health and Safety Executive, Sudbury, UK (ISBN: 978-0-71762-737-3) (out of print)

HSE (2009b) *Chemical warehousing: the storage of packaged dangerous substances*, HSG71, Health and Safety Executive, Sudbury, UK (ISBN: 978-0-71766-237-1)

HSE (2009c) *The safe use of vehicles on construction sites: a guide for clients, designers, contractors, managers and workers involved with construction transport*, HSG144, Health and Safety Executive, Sudbury, UK (ISBN: 978-0-71766-291-3). Go to: **www.hse.gov.uk/pubns/priced/hsg144.pdf**

HSE (2011a) *Workplace exposure limits. Containing the list of workplace exposure limits for use with the Control of Substances Hazardous to Health Regulations (as amended)*, EH40, Health and Safety Executive, Sudbury, UK (ISBN: 978-0-71766-446-71). Go to: **www.hse.gov.uk/pUbns/priced/eh40.pdf**

HSE (2011b) *Controlling airborne contaminants at work. A guide to local exhaust ventilation (LEV)*, HSG258, Health and Safety Executive, Sudbury, UK (ISBN: 978-0-71766-415-3).Go to: **http://tinyurl.com/qbtp8zs**

HSE (2011c) *Workplace exposure limits. Containing the list of workplace exposure limits for use with the Control of Substances Hazardous to Health Regulations (as amended), second edition*, Health and Safety Executive, Sudbury, UK (ISBN: 978-0-71766-446-7). Go to: **www.hse.gov.uk/pUbns/priced/eh40.pdf**

HSE (2012a) *Noise a work. A brief guide to controlling the risks*, INDG362(rev2), Health and Safety Executive, Sudbury, UK. Go to: **www.hse.gov.uk/pubns/indg362.pdf**

HSE (2012b) *Control the risks from hand-arm vibration*, INDG175(rev3), Health and Safety Executive, Sudbury, UK. Go to: **www.hse.gov.uk/pubns/indg175.pdf**

HSE (2012c) *Managing health and safety in zoos*, HSG219(Web15), Health and Safety Executive, Sudbury, UK (ISBN: 978-0-71762-058-6). Go to: **www.hse.gov.uk/pUbns/priced/hsg219.pdf**

HSE (2012d) *Preventing slips and trips at work*, INDG225, Health and Safety Executive, Sudbury, UK (ISBN: 978-071766-484-9). Go to: **www.hse.gov.uk/pubns/indg225.pdf**

HSE (2013a) *Workplace health, safety and welfare. Workplace (Health, Safety and Welfare) Regulations 1992. Approved code of practice and guidance*, L24, Health and Safety Executive, Sudbury, UK (ISBN: 978-0-71766-583-9)

HSE (2013b) *Heat stress in the workplace. A brief guide*, GEIS1 INDG451, Health and Safety Executive, Sudbury, UK (ISBN: 978-0-71766-468-9)

HSE (2013c) *The Control of Substances Hazardous to Health Regulations 2002 (as amended). Approved code of practice and guidance*, L5, Health and Safety Executive, Sudbury, UK (ISBN: 978-0-71766-468-9)

HSE (2013d) *Control of Asbestos Regulations 2012. Approved code of practice and guidance*, L143, Health and Safety Executive, Sudbury, UK (ISBN: 978-0-71766-618-8)

HSE (2013e) *Gas Safety (Installation and Use) Regulations 1998. Approved code of practice*, L56, Health and Safety Executive, Sudbury, UK (ISBN: 978-0-71766-617-1)

HSE (2013f) *Dust: general principles of protection, fourth edition*, EH44, Health and Safety Executive, Sudbury, UK. Go to: www.hse.gov.uk/pubns/eh44.pdf

HSE (2013g) *Legionnaire's disease: the control of legionella bacteria in water systems. Approved code of practice*, L8, Health and Safety Executive, Sudbury, UK (ISBN: 978-0-71766-615-7

HSE (2013h) *Electricity at work: safe working practices*, HSG85, Health and Safety Executive, Sudbury, UK (ISBN: 978-0-71766-581-5)

HSE (2013i) *Dangerous substances and explosive Atmospheres Regulations 2002. Approved code of practice and guidance*, L138, Health and Safety Executive, Sudbury, UK (ISBN: 978-0-71766-616-4)

HSE (2013j) *Working with display screen equipment (DSE)*, INDG36(rev4), Health and Safety Executive, Sudbury, UK. Go to: **www.hse.gov.uk/pubns/indg36.pdf**

HSE (2013k) *Ergonomics and human factors at work*, INDG90(rev3), Health and Safety Executive, Sudbury, UK. Go to: **www.hse.gov.uk/pubns/indg90.pdf**

HSE (2013l) *Workplace transport safety. A brief guide*, INDG199(rev2), Health and Safety Executive, Sudbury, UK. Go to: **www.hse.gov.uk/pubns/indg199.pdf**

HSE (2013m) *Confined spaces. A brief guide to working safely*, HSE INDG258(rev1), Health and Safety Executive, Sudbury, UK. Go to: **www.hse.gov.uk/pubns/indg258.pdf**

HSE (2013n) *Controlling fire and explosion risks in the workplace*, INDG370(rev1), Health and Safety Executive, Sudbury, UK (ISBN: 978-0-71766-485-6). Go to: **www.hse.gov.uk/pubns/indg370.pdf**

HSE (2013o) *Construction dust*, CIS36(2), Health and Safety Executive Sudbury, UK. Go to: **www.hse.gov.uk/pubns/cis36.pdf**

HSE (2014a) *Confined Spaces Regulations 1997. Approved code of practice. Regulations and guidance*, L101, Health and Safety Executive, Sudbury, UK (ISBN: 978-0-71766-622-5)

HSE (2014b) *Safe use of lifting equipment. Lifting Operations and Lifting Equipment Regulations 1998. Approved code of practice and guidance, second edition*, L113, Health and Safety Executive, Sudbury, UK (ISBN: 978-0-71766-586-0). Go to: **www.hse.gov.uk/pUbns/priced/l113.pdf**

HSE (2014c) *Working safely with solvents*, INDG273(rev1), Health and Safety Executive, Sudbury, UK. Go to: **www.hse.gov.uk/pubns/indg273.pdf**

HSE (2014d) *Safety in pressure systems*, L122, Health and Safety Executive, Sudbury, UK (ISBN: 978-0-71766-644-7). Go to: **www.hse.gov.uk/pubns/priced/l122.pdf**

HSE (2014e) *Working at height. A brief guide*, INDG401, Health and Safety Executive, Sudbury, UK (ISBN: 978-0-71766-490-0). Go to: **www.hse.gov.uk/pubns/indg401.pdf**

HSE (2014f) *Managing crowds safely. A guide for organisers at events and venues*, HSG154, Health and Safety Executive, Sudbury, UK. Go to: **www.hse.gov.uk/pubns/priced/hsg154.pdf**

HSE (2014g) *Safety in docks. Approved code of practice*, L148, Health and Safety Executive, Sudbury, UK (ISBN: 978-0-71766-572-3). Go to: **www.hse.gov.uk/pubns/priced/l148.pdf**

HSE (2014h) *Risk assessment. A brief guide to controlling risks in the workplace*, INDG163(rev4), Health and Safety executive, Sudbury, UK (ISBN: 978-0-71766-463-4). Go to: **www.hse.gov.uk/pubns/indg163.pdf**

HSE (2014i) *Safe use of work equipment. Provision and Use of Work Equipment Regulations 1998. Approved code of practice and guidance*, L22, Health and Safety Executive, Sudbury, UK (ISBN: 978-0-71766-619-5). Go to: **http://tinyurl.com/355ndtz**

HSE (2014j) *Safe use of power presses. Provision and Use of Work Equipment Regulations 1998 (as applied to power presses). Approved code of practice and guidance, second edition*, L112, Health and Safety Executive, Sudbury, UK (ISBN: 978-0-71766-620-1). Go to: **www.hse.gov.uk/pUbns/priced/l112.pdf**

HSE (2014k) *Safe use of woodworking machinery. Provision and Use of Work Equipment Regulations 1998 (as applied to woodworking machinery). Approved code of practice and guidance, second edition*, L114, Health and Safety Executive, Sudbury, UK (ISBN: 978-0-71766-620-1). Go to: **www.hseni.gov.uk/l114_safe_use_of_woodworking_machinery.pdf**

HSE (2014l) *Consulting workers on health and safety. Safety Representatives and Safety Committees Regulations 1977 (as amended) and Health and Safety (Consultation with Employees) Regulations 1996 (as amended. Approved codes of practice and guidance*, L146, Health and Safety Executive, London, UK (ISBN: 978-0-71766-461-0). Go to: **http://tinyurl.com/nd69jt6**

HSE (2015a) *Construction (Design and Management) Regulations 2015. Guidance on Regulations*, L153, Health and Safety Executive, Sudbury, UK (ISBN: 978-0-71766-626-3)

HSE (2015b) *Safety signs and signals. The Health and Safety (Safety Signs and Signals) Regulations 1996. Guidance on Regulations, third edition*, L64, Health and Safety Executive, Sudbury, UK (ISBN: 978-20-71766-598-3). Go to: **www.hse.gov.uk/pubns/priced/l64.pdf**

HSE and CBI (2009) *Guidance for electrical installation and equipment within explosives manufacturing and storage facilities including fireworks*, Health and Safety Executive , Bootle and Confederation of British Industry, London, UK (ISBN: 978-0-85201-722-7). Go to: **http://tinyurl.com/pygl5tw**

ICHEME (2004-2012) *BP Process safety series*, Institution of Chemical Engineers, Rugby, UK (ISBN: 978-0-85295-521-5). Go to: **http://tinyurl.com/p3gwq8j**

IDDON, J and CARPENTER, J (2009) *Safe access for maintenance and repair. Guidance for designers, second edition*, C686, CIRIA, London (ISBN: 978-0-86017-686-2). Go to: **www.ciria.org**

ISTRUCTE (2010) *Practical guide to structural robustness and disproportionate collapse in buildings*, The Institution of Structural Engineers, London, UK (ISBN: 978-1-90633-517-5)

JACKSON, B A, PETERSON, D J, BARTIS, J T, LATOURRETTE, T, BRAHMAKULAM, I T, HOUSER, A and SOLLINGER,M J M (2002) *Protecting emergency responders. Lessons learned from terrorist attacks*, RAND Corporation, California, USA (ISBN: 0-83303-149-X). Go to: **http://tinyurl.com/oah8zuu**

KEILLER, A, WALKER, A, LEDBETTER, S and WOLMUTH, W (2005) *Guidance on glazing at height*, C632, CIRIA, London, UK (ISBN: 978-0-86017-632-9). Go to: **www.ciria.org**

LAZARUS, D PERKINS, C AND CARPENTER, J (2006) *Safer surfaces to work on – reducing the risk of slipping*, C652, CIRIA, London, UK (ISBN: 978-0-86017-652-7). Go to: **www.ciria.org**

LONSDALE, D (2013) *The principles of tree-hazard assessment and management, Research for amenity trees, version 7*, The Stationery Office, London, UK (ISBN: 978-0-11753-355-4)

MATHER, J and LINES, I G (1999) *Assessing the risk from gasoline pipelines in the UK based on a review of historical experience*, CRR 210, Health and Safety Executive, Sudbury, UK (ISBN: 0-71761-691-6). Go to: **http://tinyurl.com/oksuka5**

NEWTON, J, NICHOLSON, B and SAUNDERS, R (2011) *Working with wildlife: guidance for the construction industry*, C691, CIRIA, London, UK (ISBN: 978-0-86017-691-6). Go to: **www.ciria.org**

NHBC (2011) *The Building Regulations 2004 edition – England and Wales: requirement A3 – disproportionate collapse. Technical guidance note*, National House Building Council, Milton Keynes, UK. Go to: **http://tinyurl.com/ndz7fbz**

OVE ARUP (2015) *CDM 2015 – construction work sector guidance for designers, fourth edition*, C755, CIRIA, London (ISBN: 978-0-86017-756-2). Go to: **www.ciria.org**

PAGE, M and HOUGH, A (1989) *Ergonomic aspects of escalators in retail organisations*, CRR12, Health and Safety Executive, Sudbury, UK. Go to: **www.hse.gov.uk/Research/crr_pdf/1989/crr89012.pdf**

RAW, G (2001) *Building Regulation health and safety*, BR417, BRE Press, Building Research Establishment, Garston, UK (ISBN: 1-86081-475-1)

ROAD LIAISON GROUP (2004) *Well-lit highways. code of practice for highway lighting management*, UK Lighting Board, The Stationery Office, Norwich, UK (ISBN:L 0-11552-632-3). Go to: **http://tinyurl.com/pb8qqt4**

ROWE, I and ANCLIFFE, S (2009) *Guidance on designing for crowds: an integrated approach*, C675, CIRIA, London, UK (ISBN: 978-0-86017-675-6). Go to: **www.ciria.org**

RSSB (2009) *Crowd management on trains: a good practice guide*, T605, Rail Safety and Standards Board, London, UK. Go to: **http://tinyurl.com/nwcamgu**

SCADDAN, B (2011) *IEE wiring regulations – explained and illustrated, 17th edition*, Elsevier Ltd, MA, USA (ISBN: 978-0-08096-914-5)

SCARLETT, A J and STAYNER, R M (2005) *Whole-body vibration on construction, mining and quarrying machines*, RR400, Health and Safety Executive, Sudbury, UK. Go to: **www.hse.gov.uk/research/rrpdf/rr400.pdf**

STRICKLAND, C (1997) *Best practice in the specification of offices, second edition*, British Council for Offices, London, UK(ISBN: 978-0-95241-312-7)

TRADA (2005) *Wood preservation – chemicals and processes*, TRADA Wood Information Sheet Section 2/3 Sheet 33, revised, Timber Research and Development Association, High Wycombe, UK. Go to: **http://tinyurl.com/nvkk2xh**

WAY, A G J (2004) *Guidance on meeting the robustness requirements in Approved Document AP341*, Steel Construction Institute, Berkshire, UK (ISBN: 1-85942-163-6)

WILSON, S, OLIVER, S, MALLETT, H, HUTCHINGS, H and CARD, G (2007) *Assessing risks posed by hazardous gases to buildings*, C665, CIRIA, London (ISBN: 978-0-86017-665-7). Go to: **www.ciria.org**

Statutes

Acts

Equality Act 2010 (c.15)

Health and Safety at Work etc. Act 1974 (c.37)

Radioactive Substances Act 1993 (c.12)

Directives

Directive 94/9/EC of the European Parliament and the Council of 23 March 1994 on the approximation of the laws of the Member States concerning equipment and protective systems intended for use in potentially explosive atmospheres (OJ L 100, 19.4.1994) (ATEX Directive)

Regulations

Chemicals (Hazard Information and Packaging for Supply) (CHIPS) Regulations 2009 (No.716

Confined Spaces Regulations 1997 (No.1713)

Control of Asbestos Regulations 2006 (No.2739)

Control of Lead at Work Regulations 2002 (No.2676)

Control of Major Accident Hazards Regulations (COMAH) 1999 (No.743)

Control of Noise at Work Regulations 2005 (No.1643)

Control of Substances Hazardous to Health (COSHH) Regulations 2002 (No.2677)

Control of Vibration at Work Regulations 2005 (No.1093)

Dangerous Substances and Explosive Atmospheres Regulations (DSEAR) 2002 (No 2776)

Docks Regulations 1988 (No 1655)

Electricity at Work Regulations 1989 (No.635)

Equalities Act 2010 (c.15)

Gas Safety (Installation and Use) Regulations 1998

Health and Safety (Display Screen Equipment) Regulations 1992

High-Activity Sealed Radioactive Sources and Orphan Sources (HASS) Regulations 2005 (No.2686)

Ionising Radiations Regulations 1999 (No.3232)

Ionising Radiations (Medical Exposure) Regulations 2000 (No.1059)

Lifting Operations and Lifting Equipment Regulations (LOLER) 1998 (No.2307)

Management of Health and Safety at Work Regulations 1999 (No.3242)

Provision and Use of Work Equipment Regulations (PUWER) 1998 (No.2306)

Radiation (Emergency Preparedness and Public Information) Regulations 2001 (No.2975)

The Regulatory Reform (Fire Safety) Order 2005 (No 1541)

The Shipbuilding and Ship-repairing Regulations 1960 (No.1932)

Workplace (Health, Safety and Welfare) Regulations 1992 (No.3004)

The Work at Height Regulations 2005 (No. 735)

Standards

BS 4211:2005+A1:2008 *Specification for permanently fixed ladders*

BS 4592-1:2006 *Industrial type flooring and stair treads. Metal open bar gratings. Specification*

BS 5395-1:2010 *Stairs. Code of practice for the design of stairs with straight flights and winders*

BS 5839-1:2013 *Fire detection and fire alarm systems for buildings. Code of practice for design, installation, commissioning and maintenance of systems in non-domestic premises*

BS 5656-2:2004 *Escalator and moving walks. Safety rules for the construction and installation of escalators and moving walks. Code of practice for the selection, installation and location of new escalators and moving walks*

BS 5958-1:1991 *Code of practice for control of undesirable static electricity. General considerations*

BS 6180:2011 *Barriers in and about buildings. Code of practice*

BS 6465-1:2006+A1:2009 *Sanitary installations. Code of practice for the design of sanitary facilities and scales of provision of sanitary and associated appliances*

BS 7671:2008+A3:2015 *Requirements for electrical installations: IET wiring regulations*

BS 7915:1998 *Ergonomics of the thermal environment – guide to design*

BS 8213-1:2004 *Windows, doors and rooflight. Design for safety in use and during cleaning of windows, including door-height windows and roof windows. Code of practice*

BS 8300:2009+A1:2010 *Design of buildings and their approaches to meet the needs to disabled people. Code of practice*

BS 8437:2005+A1:2012 *Code of practice for selection, use and maintenance of personal fall protection systems and equipment for use in the workplace*

BS 9999:2008 *Code of practice for fire safety in the design, management and use of buildings*

BS EN 12101-6:2005 *Smoke and heat control systems. Specification for pressure differential systems. Kits*

Websites

Association for Project Safety (APS): **www.aps.org.uk**

BRE Radon: **www.bre.co.uk/page.jsp?id=3133**

Constructing Better Health (CBH) Scheme: **www.cbhscheme.com**

Public Health England Radon: **www.ukradon.org/information**

Structural-Safety: **www.structural-safety.org**

Gov.uk

General guidance: **https://www.gov.uk**

Land contamination technical guidance: **http://tinyurl.com/nk6xk84**

Radiation: **https://www.gov.uk/health-protection/radiation**

HSE

Catering and hospitality: **www.hse.gov.uk/catering**

Display screen equipment: **www.hse.gov.uk/msd/dse**

Roadways/site traffic control/immobilisation of vehicles: **www.hse.gov.uk/comah/sragtech/techmeastraffic.htm**

A1 Risk checklist

The risks groups and topics may be adopted in a 'risk checklist' to facilitate discussion of a particular project.

Potential risks	Project notes
Physical environment • lighting • noise • vibration • temperature • wetness and humidity • draughts • confined spaces. ***Chemical/biological environment*** • sanitary conditions • animals • moulds and fungal growths • hazardous materials (solids/liquids/gases/fumes) • smoke • dusts and fibres • contamination/pollution. ***Hazardous systems*** • electricity • hot water and steam • piped gases/liquids • hot/cold surfaces • storage • radiation. ***Normal activities*** • posture and manual handling • use of vehicles • use of plant and equipment • industrial processes • working on highways • use of doors and windows/glazing • use of lifts, escalators and moving walkways. ***Slips and trips*** • while in motion on floors and ramps • while using stairs or escalators • while static. ***Working at height*** • using ladders etc • at unprotected edges/fragile surfaces. ***Abnormal events*** • fire • explosion • falling objects • disproportionate collapse • drowning and asphyxiation • over-crowding • responding to emergency events • malicious human intervention. ***Anything else?***	

A2 Example of the use of a risk checklist to record team discussions

This example is for a small shelter for taxi drivers to be located alongside a busy street. Inside the shelter there will be facilities for making tea/coffee and hot snacks. The example records the consideration of risks by the design team at their first design team meeting. The checklist is used to record the important points discussed and decisions made.

Risk	Risk elimination and reduction	Actions
Physical environment	The shelter will comply with Building Regulations and will be able to accommodate up to 20 people in a small space. There will be a lobby where wet clothes can be left to dry. The shelter will be maintained as a warm dry space 24/7.	
Chemical/biological environment	Users will eat in the shelter and may or may not use the toilet.	Provide hand-wash facility at entrance/exit as well as in toilet.
Hazardous systems	Shelter may be left empty for long periods, so an open fire or naked flames would be an unacceptable risk	All heating (space, water) will be by electricity (no bottled gas). Food heated by microwave oven.
Normal activities	The shelter will comply with Building Regulations.	Chemical toilet to be in a separate naturally-ventilated space.
Slips and trips	Wet shoes and leaves etc.	Flooring should be low slip-risk material.
Working at height	Single-storey roof to be sloped and self-cleaning finish.	
Abnormal events	Overhanging trees may shed branches. Risk of smokers starting a fire. Risk of impact from traffic. Sited in a busy well-lit area. However, there is some risk of fires being started.	Suggest local authority to be asked to check trees annually. Investigate the best way to manage this risk (see *Notes*). A heavy duty kerb and guard-rail will be provided. Structure to be fire resistant. Provide no smoking signs, extinguishers, smoke detector.
Anything else?	Nothing else identified.	

Notes

Risk of smokers starting a fire was identified as a serious issue. However, after discussion it was considered most unlikely that this would be a safety issue unless:

1 Materials used were highly flammable such that the fire could develop and release toxic fumes before escape was possible.

2 There was no-one present, in which case there would only be loss of property. The design team decided to specify fire-resistant materials and to provide a fire extinguisher. Also, after consultation with users, smoking inside the shelter was banned.

A3 Sources of further information

Legislation and approved codes of practice (ACOPs)	
Regulation	***How it helps***
Building Regulations comprise the following parts: Part A Structure Part B Fire safety Part C Site preparation and resistance to moisture Part D Toxic substances Part E Resistance to the passage of sound Part F Ventilation Part G Hygiene Part H Drainage and waste disposal Part J Combustion appliances and fuel storage systems Part K Protection from falling and impact Part L Fuel conservation Part M Access to and use of buildings Part N Glazing safety Part P Electrical safety – dwellings	The Regulations were conceived primarily for housing standards. The Regulations set out the requirements. Each part has its own Approved Documents setting out approved methods for dealing with the requirements and it is these documents that are normally referred to
The Health & Safety at Work etc Act 1974	The 'head' regulation requiring that workplaces are to be regulated and setting out responsibilities
The Construction (Design and Management) Regulations 2015 HSE (2015a) L153	CDM 2015
Workplace (Health, Safety and Welfare) Regulations 1992 HSE (2013a) L24	Sets out specific requirements for workplaces (see Appendix A2 for a summary)
The Management of Health and Safety at Work Regulations 1999 L21 (HSE, 2000f)	Sets the requirements for those who manage a workplace
Equalities Act 2010	Places obligations on building owners, employers and others to ensure that they do not discriminate against people's
The Provision and Use of Work Equipment Regulations (PUWER) 1998 HSE (2014i) L22 HSE (2014j) L112 HSE (2014k) L114	
The Lifting Operations and Lifting Equipment (LOLER) Regulations 1998 HSE (2014b) L113	
The Personal Protective Equipment at Work Regulations 1992 HSE (2005d) L25	Regulations for PPE to be used in a workplace
The Health and Safety (Display Screen Equipment) Regulations 1992 as amended 2002 HSE (2003a) L26	Regulations for the use of computers in a workplace
The Manual Handling Operations Regulations 1992 as amended 2002 HSE (2004) L23	Regulations for manual handling
The Control of Substances Hazardous to Health (COSHH) Regulations 2002 HSE (2013c) L5	

Legislation and approved codes of practice (ACOPs)	
Regulation	***How it helps***
The Chemical (Hazard Information and Packaging for Supply) (CHIP) Regulations 2002 (as amended 2005)	Requiring that suppliers make available data sheets setting out the hazards of materials and products
Control of Asbestos Regulations 2006 HSE (2013d) L143	Regulations for dealing with asbestos
Control and Use of Lead at Work Regulations 2002 HSE (2002a) L132	Regulations for dealing with lead
The Ionising Radiation Regulations 1999 HSE (2000b) L121	Regulations for dealing with radioactive materials
HSE (2013g) L8	Advises how to manage the risks of Legionnaire's Disease
Electricity at Work Regulations 1989	Regulations for electricity
Gas Safety (Installation and Use) Regulations 1998 HSE (2013e) L56	Regulations for gas
Dangerous Substances and Explosive Atmospheres Regulations 2002 and a set of ACOPs as follows: L138 Dangerous substances and explosive atmospheres.	Regulations and guidance for dangerous substances and explosive atmospheres
The Chemical (Hazard Information and Packaging for Supply) (CHIP) Regulation 2002 (as amended 2005)	Requiring that suppliers make available data sheets setting out the hazards of materials and products
Control of Asbestos Regulations 2006 HSE (2013d) L143	Regulations for dealing with asbestos
Control and Use of Lead at Work Regulations 2002 HSE (2002a) L132.	Regulations for dealing with lead
The Control of Major Accident Hazards Regulations (COMAH) 1999 HSE (2006e) L111.	
Safety of pressure systems. Pressure Safety Systems Regulations 2000 HSE (2014d) L122	Regulations for pressure systems
Safe Work in Confined Spaces Regulations 1997 HSE (2014a) L101.	Regulations for work in confined spaces
The Regulatory Reform (Fire Safety) Order 2005	Relates to fire assessment of the workplace
The Work at Height Regulations 2005	Regulations for working 'at height'
The Control of Vibration at Work Regulations 2005 HSE (2005a) L140 HSE (2005b) L141	Regulations for vibrations that affect people
The Control of Noise at Work Regulations 2005 HSE (2005c) L108.	Regulations for noises that affect workers
Health and Safety (Consultation with Employees) Regulations 1996 HSE (2014l) L146.	Requires consultation with non-unionised employees
The Health and Safety (Safety Signs and Signals) Regulations 1996 HSE (2015b) L64	Regulations for safety signs etc
The Work in Compressed Air Regulations 1996 HSE (1996c) L96	Mainly applies to tunnelling but there may be work processes where work in compressed air is necessary
The Ionising Radiations Regulations 1999 HSE (2000b) L121	Regulations for ionising radiation
The Ionising Radiations (Medical Exposure) Regulations 2000	Regulations for ionising radiation in medical work
The Radioactive Substances Act (1993)	Regulations principally concerning waste

Legislation and approved codes of practice (ACOPs)	
Regulation	*How it helps*
High Activity Sealed Radiation Sources and Orphan Sources (HASS) Regulations 2005	Principally concerning nuclear activity
The Radiation (Emergency Procedures and Public Information (REPPIR) Regulations 2005	
Dangerous Substances and Explosive Atmospheres Regulations 2002 HSE (2013i) L138	Regulations applicable to many workplaces

Notes

1 Where possible, references are included in Chapter 3 tables. This appendix contains a list of relevant legislation together with references of a more general nature.

2 Where reference is made to an ACOP this refers to an approved code of practice in support of particular legislation. Some ACOPs contain guidance material that has a different legal significance.

Regulations applicable to specific industries	
HSE (1980) HG7	Container terminals
Docks Regulations 1988 HSE (2014g) L148	Docks
HSE (1995c) L60	Potteries
Shipbuilding and Ship-repairing Regulations 1960	Shipbuilding and repair

HSE guidance (note that further guidance is available through the HSE website, including downloadable simple guides. Search on topic words as listed there)	
Guidance	*How it helps*
HSE (2015a) L153	Gives HSE advice on CDM 2015
HSE (2014h) INDG163(rev4)	Explains the basics of risk assessment.
HSE (2006f)	Explores the subject of occupational health.
HSE website	Advice particular to catering facilities.

Other reference material	
Guidance	*How it helps*
BS 8300:2009+A1:2010	A code of practice to assist detailed design.
BS 8437:2005+A1:2012	A code of practice for personal fall protection systems.
Hawkins (2011)	Sets out typical requirements for different spaces and suggests allowances to make at scheme design stage.
CIBSE guides: Guide A Environmental design Guide B Heating, ventilation, air conditioning and refrigeration Guide D Transportation systems in buildings Guide E Fire engineering Guide G Public health and plumbing engineering Guide K Electricity.	These design guides provide design information and calculation techniques. They are continually under review and update.
CIRIA (1999) C507	Guidance on O&M manuals.
Armstrong (2002) C581	Guidance on FM manuals.
HSE catering information sheets	A series of sheets giving advice on issues in the catering world, covering the key inherent risks.
HSE (1999) HSG129	Guidance for engineering workshops.
Raw (2001) BRE Report 417	Analysis of in-use health and safety risks, including discussion of medical conditions.
Structural-Safety (various reports)	Advice about issues of concern reported by professionals.